ENCYCLOPAEDIA OF CELL BIOLOGY - 5

CYTOCHEMISTRY

By

Dr. M. Prakash

Dept. of Zoology
M.M.H. Post Graduate College
Ghaziabad (U.P.)
(India)

D P H

DISCOVERY PUBLISHING HOUSE PVT. LTD.

NEW DELHI-110 002

Reprinted: 2014

ISBN 978-81-8356-563-9

Published by:

DISCOVERY PUBLISHING HOUSE PVT. LTD.

4831/24, Ansari Road, Prahlad Street
Darya Ganj, New Delhi-110002 (India)
Phone: 23279245, 43764432 • Fax: 91-11-23253475
E-mail: parul.wasan@gmail.com
info@discoverypublishinggroup.com
Website: www.discoverypublishinggroup.com

Printed at: Dynamic printers, Delhi

Preface

The present title *"Cytochemistry"* has been written for undergraduate and postgraduate students of all Indian universities. It is a text on structural and functional unit of the organism that has been thoroughly checked and ornamented with easy and clear illustrations. Special emphasis has been concentrated on the fundamentals of the subject. All related students will find this well established text eminently suitable for introductory courses. Professional cell biologists, researchers, and students may find it useful for updating, and reference purposes. One of our major concerns of this title is to provide students with a basic understanding of what a cell is and why the cell is the fundamental unit of life. Our approach throughout has been to focus on the major question involved and the experimental approaches utilized in addressing these questions. Each important point is illustrated with an example to help clarify the idea; the numerous line drawings have all been chosen or created to help convey particular points. The result is a text that is easy to read, without resorting to verbal gimmicks or talking down to readers.

To make the work more comprehensive and informative, the author has consulted many authoritative books, research journals, abstracts, monographs etc., so there can be no claim to originality except in the manner of treatment.

The author expresses his thanks to his friends and colleagues whose continue inspirations have initiated him to bring out this book.

The author expresses his gratitude to Mr. Wasan and staff of M/s Discovery Publishing House Pvt. Ltd. for their whole hearted cooperation in the publication of this book.

Author

Preface

Contents

1

Chemistry of Protoplasm

Protoplasm is a highly complex mixture of some elements and compounds found in the bodies of living beings. The protoplasm is variously known as the *living matter*, *living substance* or the *physical basis of life*. It is the basic fundamental substance exhibiting all the vital processes of the cell.

The protoplasm was first observed by *Corti*. *Dujardin* a Frenchman described it as a soft and gumy substance and named it as *sarcode i.e. flesh*. *Purkinje*, a Bohemian physiologist, was the first biologist, who gave the name *protoplasm* to this living substance. *Hugo von Mohl*, a German botanist, also suggested the name protoplasm for the granular and viscous substance found in plants similar to that found in the animals.

The above-mentioned German botanist popularised the word protoplasm as a name given to the living matter found in plants and animals. Protoplasm is the most complex. and interesting substance. It is not to be thought of a chemical compound but rather as very complex organised system.

Protoplasm varies somewhat in its nature from cell to cell and. from organism to organism, but basically it must be the same, as evidenced by its common manifestations of metabolism, growth, reproduction and by some other peculiarities.

PHYSICAL NATURE OF PROTOPLASM

Different workers have proposed different theories to explain the physical nature of protoplasm as given under the following heads:

Granular Theory

This theory was propounded by *Altmann*. According to this theory, protoplasm consists of numerous tiny granules as shown in *Amoeba*. *Henle*, *Maggi*, *etc*., considered these protoplasmic granules as *plastidules*. *Altmann* recognised them as "elementary organisms", or *bioplasts* (or cytoplasts).

Alveolar Theory

The alveolar nature of protoplasm was suggested by *Butschilii* in 1892. According to him, protoplasm consists of many suspended droplets or alveoli or minute bubbles, resembling the foams of emulsion.

Fibrillar Theory

This theory was put forward by *Flemming*. According to him, protoplasm consists of fibres embedded in the inner mass of matrix. The fibrillae are called mitome or spongioplasm and ground substance is termed paramitome or hyaloplasm.

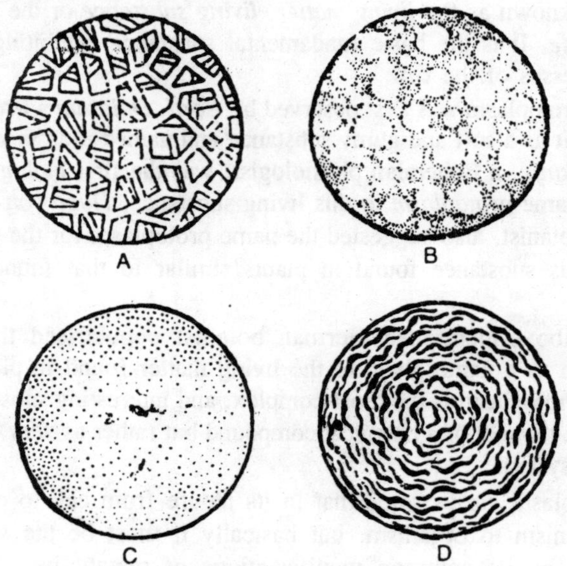

Figure 1.1 : Physical appearance of protoplasm.
A. Reticular B. Alveolar C. Granular D. Fibrillar.

Reticulate Theory

This theory was proposed by *Klein*, *Cornoy etc.* suggesting that the protoplasm consists of a network or reticulum of fibres in the ground substance.

Colloidal Theory

Proposed by *Wilson* in 1925. According to this theory the protoplasm having any of the aforesaid appearances is always a fluid-colloidal system having various chemical inclusions in *gel phase*. A gel is a semi-solid condition which presents jellylike appearance. Here the molecules are held together by various bonds depends upon their nature and strength. By absorbing water the gel changes to more liquid-like phase.

This is known as *sol* and the process is termed as *solation*. The sol can stream easily and by losing water again changes into gel-state. Both these states of colloidal matrix are interchangeable according to the various physiological, mechanical and biochemical activities of the cell. This represents phase reversal in the colloidal system. These sol-gel conditions of colloidal system are the basis for the mechanical behaviour of protoplasm.

PROPERTIES OF PROTOPLASM

Cohesiveness

The various particles or molecules of protoplasm are adhered with each other by forces, such as Van der Waal's bonds, that hold long chains of molecules together. These Van der Waal's bonds are weak and non-specific forces between non-polar groups of atoms. This property varies with the strength of these forces.

Contractility

This property is significant in various stomatal operations in plants. The contractility of protoplasm is important for the absorption and removal of water as they generally occur in protoplasm.

Electrical Charge

Protein molecules repel each other because their overall' charge is similar. All molecules are either positively charged or negatively charged. However, if the molecules approach one another close enough so that valency forces can act, then they may be attracted to each other.

Precipitation

Addition of certain amount of electrolyte in a colloidal' suspension causes its dispersed particles to colloide, aggregate and finally to precipitate as suspension. For example, the addition of HCl to a colloidal system of arsenic sulphide causes precipitation.

Viscosity

The viscosity of the ground substance of the cell varies greatly. It may be as low as that of water, or may be very high in the gelating cytoplasm of pseudopodia of *Amoeba*.

Streaming Movement or Cyclosis

The protoplasm exhibits various sorts of streaming movements inside the cell. These have been studied in *Amoeba*, *Paramecium etc*. The movement involves only the localised portions with no visible changes in the protoplasm. No complete explanation of this has been given as yet. But it is seen that its rate depends upon the rate of cell metabolism. It is due to the fact that the energy is supplied by respiration.

Amoeboid Movement

The amoeboid movement as exhibited by *Amoeba* and other protozoans involves the movement of entire protoplasm of the cell, where the cytoplasm moves as one mass carrying the various inclusions with it. It is due to the continuous change of gel to sol and sol to gel.

Brownian Movement

It is characterised by the zigzag motion of suspended colloidal particles, occurring due to the bombardment of one particle or molecule by other. This type of movement of particles was first of all observed by Robert Brown in 1827 in the colloidal solution and hence such movements are known as *Brownian* movements.

The higher the temperature, more rapid the movement and thus viscosity of cell is decreased. This means that high viscosity indicates a more gel-like state of protoplasm and low viscosity, a more sot-like condition.

Tyndall Effect

Colloidal particles of protoplasm have the property of scattering light. When a beam of light is passed through a colloidal solution it becomes visible. This is the Tyndall effect. A colloidal solution of proteins in water shows a typical tyndall cone.

Adsorption

The tendency of particles, molecules or ions to adhere to the surface of certain solids or liquids is known as adsorption and is exhibited by the particles of colloidal system. The phenomenon helps the matrix to form protein boundaries.

Biological Properties

Protoplasm has all the biological properties of a living organism. It is capable of nutrition, respiration, excretion, metabolism, growth and reproduction. It has the property of irritability, *e.g.* it responds to stimuli like heat, light and chemicals. It also has the property of conductivity, *i.e.* of conducting impulses produced by stimuli.

CHEMICAL NATURE OF PROTOPLASM

For the detailed study the chemical components of the cell can be classified as inorganic (mineral salts) and organic (proteins, carbohydrates, nucleic acids, lipids and so forth) substance. Although the most prominent costituent of protoplasm is water-the substance which gives protoplasm its characteristic structure is protein.

Lipids are important in all membranes and carbohydrates serve as nutrient stores. The protoplasm of a plant or animal cell contains 75 to 85% water, 10-20% protein, 2-3% lipids, 1% carbohydrats and 1% inorganic material. The following table elsewhere in this chapter gives approximate figures of the relative amounts of the main inorganic and organic compounds found in active protoplasm.

Table 1.1.

Substance	*Percent*	*Average molecular weight*	*Number of molecules in relation to protein*
Water	85	18	180
Protein	10	36000	1
DNA	0.4	10°	—
RNA	0.7	4.0×10^4	—
Lipid	2	700	10
other organic matter	0.4	250	20
Inorganic substances	1.5	55	100

Water

Water is the main component of protoplasm and occurs in large amount, about 85%. It serves as a natural solvent for other materials and also plays a major role in the metabolic activities, since physiologic processes occur exclusively in an aqueous medium. The water molecules also participate in many enzymatic reactions in the cell.

No doubt, certain amount of water is formed in the cell as a result of metabolic processes, but this is insufficient to maintain the water balance of the body. Therefore, it must be and is supplied from the outside. Inside the cell, the water exists free, *i.e.* a solvent as well as bound, i.e. tied to polar groups of protein molecules by hydrogen bonds.

INORGANIC COMPOUNDS

A number of inorganic salts occur both in free as well as in the ionised state. The elements that occur in quantity in protoplasm are oxygen, carbon, hydrogen, and nitrogen. Always present but in much smaller amounts are potassium, phosphorus, calcium, sulphur. magnesium and iron. These ten are generally referred to as *essential elements*. Chlorine and sodium are necessary components of most animal protoplasm but are apparently not essential for plant protoplasm.

Copper, boron, iodine, maganese, zinc and several other elements which are present in all protoplasm but only in minute quantities, are called *trace elements*. This term must be used with care, because these elements are still important even though they are present only in very small quantities. The essential elements found in human protoplasm are listed below

Table 1.2. Showing percentage by weight to the elements in protoplasm.

Elements	*Symbol*	*Percent*
Oxygen	O	65.04
Carton	C	18.24
Hydrogen	H	10 05
Nitrogen	N	3.15
Potassium	K	1.60
Phosphorus	P	0.84
Calcium	Ca	0.25
Sulphur	S	0.20
Magnesium	Mg	0.04
Iron	Fe	0.01
Chlorine	Cl	0 27
Sodium	Na	0.26

The elements are almost always present in protoplasm in the form of chemical compounds rather than elements. Many of these compounds are inorganic salts, which are usually in solution. These salts have numerous functions.

They serve as sources of elements to be built into other compounds, and some act as buffer in maintaining the proper acidity or alkalinity in the protoplasm. Because salts in solutions are electrical conductors, they also function in some way in connection with the electrical properties of protoplasm, however, this is not well understood at present.

ORGANIC COMPOUND

Proteins

Proteins are the framework of protoplasm. About 10;1. of protoplasm is protein. Proteins are linear polymers of high molecular weight. All proteins contain C, H, O and N, the presence of N distinguishing them from carbohydrates and fats. On an average proteins contain 16% nitrogen. Some proteins also have S in addition, and in a few proteins P and other elements may be present.

The molecular weight of proteins varies from about 12,000 daltons (bovine insulin) to several million. Cells contain a very large number of proteins. The number may vary from 1000-2000 in the simplest bacteria to as many as 100,000 different proteins in human cells.

Protein are found in two form: Simple and conjugate.

The Simple Proteins

Are compounds, which on hydrolysis, yield exclusively alpha *(α)* amino-acids. In this group the most important proteins are the albumins (soluble in water and coaguable by heat); the globulins (insoluble in water, but soluble in dilute salt solutions); the protamines, strongly alkaline components of the sperm cell (e.g. clupein, salmine, etc.); and the some what less basic histones which are found in many cell nuclei.

Conjugate proteins are a combination of a simple protein and another substance called the *prosthetic group*. Unlike the simple protein, the conjugate protein yields on hydrolysis alpha *(α)* amino-acids and an organic component. Nucleoproteins are also included in conjugate protein group which play an important role in the cell and whose prosthetic groups are the nucleic acids; the *glyco-protein* (mucoproteins) in which protein is combined with a carbohydrate; the *lipo-proteins*, which are combinations of proteins with higher fatty acids; and the chromoproteins, a very widely distributed group, included a series of substances of great bilogical importance which are characterised by their particular colours.

Among the chromoproteins are haemoglobin, in which the globulin is combined with an iron-porphyrin compound, and the haemocyanin which occurs in the blood of various invertebrates and in which copper is found in a position similar to that of iron in haemoglobin. A series of respiratory enzymes (cytochromes, flavoprotein etc.,) belong to the chromoprotein group.

SIGNIFICANCE OF PROTEINS

Proteins are the most significant compounds in the body of living organisms because of the following physiological roles performed by them

1. *Proteins confer specific individuality to living structures, organisms and the species*. This is known as biochemical individuality-Protein individuality is well illustrated by blood transfusion experiments. The blood of one species of animals cannot be transfused into the circulation of animals of other species. Not only this, the blood proteins of different individuals of the same species are not identical.

2. *Proteins play an important role in life processes*. These can combine with both acid and bases and can neutralize excess of any one of them in the protoplasm. These check the accumulation of H_+ or OH_- ions in the protoplasm.

3. *Proteins have a high chemical potentiality* and, therefore, readily react with many substances, therefore, these act as enzymes and catalyse almost all the life processes of living organism.

4. *Proteins form boundaries of living cells* and their components and thus do not permit an easy access to and from the -cell cytoplasm.

5. *Protein differentiation of organs and animals*. How only twenty amino acids lead to the formation of thousands of proteins is a complex but interesting question. For easy understanding of this phenomenon the twenty amino acids are linked to 20 alphabets of English language and the formation of a protein with the construction of a word.

For example, an insulin molecule formed of 16 amino acid molecules is similar to a word formed of sixteen letters. In English dictionary 26 alphabets combine to form at least 2,00,000 words. Similarly 20 amino acids can give approximately same number or even more varieties of proteins. Such a wide variety of proteins will naturally bring wide range of differentiation.

AMINO ACIDS

There is a common plan of construction for the thousands of kinds of proteins in living systems. The 20 kinds of naturally-occurring amino acid monomers are strung together in unbranched, linear polymer chains of proteins.

These are the 20 amino acids specified in the genetic code that is universal to all organisms. Some other kinds of amino acids are also found in cells, but they are either degradation products or residues that have been modified from one of the 20 commonly occurring amino acids after this latter has been inserted into the polymer chain. Hydroxyproline is a major amino acid constituent of collagen in connective tissue, but proline residues are initially included in the protein and are converted to hydroxyproline after polymerisation. Hydroxyproline is not one of the encoded amino acids, but proline is. Many proteins contain fewer than 20 kinds of amino acids.

The relative proportions and the absolute numbers of the amino acid repertory vary from one protein to another, as a reflection of the specific information in genes, which are the blueprints for protein

CARBOHYDRATES

The carbohydrates (L., *carbo-carbon* or coal, *Gr*., *hydro*water) are the compounds of the carbon, hydrogen and oxygen. They form the main source of the energy for all living beings. Only green parts of plants and certain microbes have the power of synthesizing the carbohydrates from the water and CO_2 in the presence of sunlight and chlorophyll by the process of *photosynthesis*. All the animals, non-green parts of the plants (viz., stem, root, etc.), non-green plants (e.g., fungi), bacteria and viruses depend on green parts of plants for the supply of carbohydrates.

Carbohydrates can be classified into:

Monosaccharides

These are simple sugars with an empirical formula Cn $(H_2O)n$. These may be of trisoes if they have 3 carbons, tetroses, if they have four carbons, pentsoses type, if they have five carbons, or hexoses, if they have six. Glucose or grape sugar and fructose or fruit sugar are examples of bexoses type.

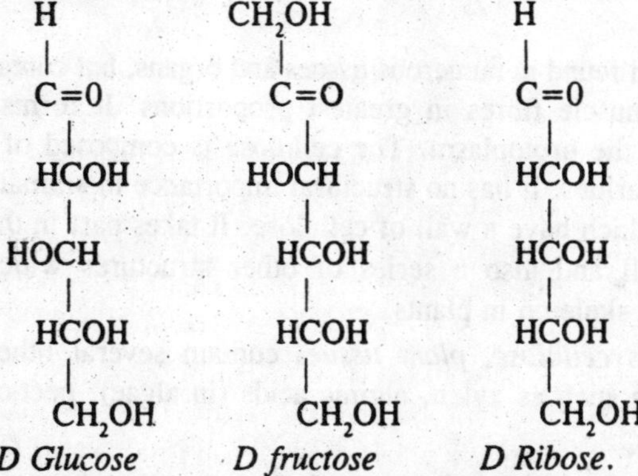

D Glucose *D fructose* *D Ribose*.

And ribose and deoxyribose (the sugars present in nucleic acids) are examples of pentose. The monosaccharides are readily soluble in water and can pass across the membrane and can also be crystalised. This soluble nature is changed, when these are stored in the animal as well as in plant tissues. In, such cases it is usually polymerised to an insoluble form such. as glycogen (animal starch) in animals and starch in plants.

Disaccharides

The disaccharides are having the empirical formula $C_{12}H_{22}O_{11}$ and are readily soluble in water and these can also be crystalised and passed through the membrane. The commonest and best known disaccharide is sucrose or table sugars which is stored in such plants as the sugarcane and the sugar beet. On hydrolysis yields monosaccharides or simple sugars.

Polysaccharides

These are formed by the condensation of many molecules of monosaccharides with a corresponding loss of water molecules. Their empirical formula is $(C_{16}H_{10}O_5)$n. They yield molecules of simple sugars on hydrolysis. Biologically important polysaccharides are the starch, the glycogen, and the cellulose. The former two are the reserve substances in cells of plants and animals respectively where as the cellulose is the characteristic structural element of the plant cells.

The starch $(C_6C_{10}O_5)$n is a mixture of two long polymer molecules, i.e., linear amylase and branched amylopectin. It has no structural importance in animals. The glycogen $(C_6H_{12}O_6)$n is generally called as the starch of animal cells. It is a polymer composed of many molecules of glucose and thus represents an important reserve of energy in the body.

Though found in numerous tissues and organs, but contained in liver cells and muscle fibres in greatest proportions. It forms a colloidal solution in the protoplasm. The cellulose is composed of hundred of monosaccharides. It has no structural importance in animals except the tunicates which have a wall of cellulose. It takes part in the formation of cell wall and also a series of other structures which form the suipporting skeleton in plants.

Besides *cellulose*, *plant tissues* contain several other structural components such as xylen, alginic acids (in algae), pectic acid, etc.

Significance

Thus we can conclude that carbohydrates play three main roles in the cells. These are

1. They furnish the most readily available fuel.
2. They are important articles of storage.

3. They have a minor role in furnishing part of the structural material of atleast one part of the essential environment of living cells.

LIPIDS

Lipids comprise a rather heterogeneous class of compounds which are sparingly soluble in water, but show considerable solubility in organic solvents like either, chloroform, benzene, hot alcohol and petroleum ether. Their physical properties indicate the hydrophobic nature of the hydrocarbon structure, although lipids may also contain some hydrophilic groups.

Lipids have several biological functions.

(1) Certain lipids e.g. triglycerides are storage compounds for the reserve energy of the body, e.g. lipids of the fat body and subcutaneous fat.

(2) Lipids are important components of cell membranes. The plasma membrane of eukaryote cells is composed primarily of lipids and proteins, along with carbohydrates in some cases.

(3) In nerve fibers the myelin sheath contains linids which act as electrical insulators.

(4) As components of subcutaneous tissue lipids are important for insulation of heat.

(5) Some hormones *e.g.* steroids are lipids.

(6) Some lipids are important as vitamins.

(7) Lipids also occur as components of some enzyme systems.

Lipids can be classified as follows

(I) Simple lipids, (II) Compound lipids.

Simple Lipids

Simple lipids are the esters of the alcohols-or the trigl5 cerides containing fatty acids and alcohols.

$$\text{Triglyceride} \xrightarrow[H_2O]{\text{Lipase}} \text{Glycerol} + 3 \text{ Fatty acids}$$

The constituent fatty acids may be *saturated*, *e.g.*, palmatic, stearic acids, or *unsaturated*, *e.g.*, oleic, linoleic, arachedonic and clupanadonic acids. The lipids of the matrix of animal cells contain fatty acids, viz., palamatic, stearic palmitoleic, oleic, linoleic and inolenic acids. The simple lipids of the matrix are as follows

Natural fats

The natural fats are the naturally occurring fats which occur in animal and plant cells as stored food substances.

Waxes

Waxes are esters of higher fatty acid with alcohols other than glycerols. In the human body the commonest waxes are the cholesterol esters. They are most abundant in the blood, suprarenal glands, the gonads and the sebaceous glands of the skin.

The natural waxes are solid at ordinary temperature and are not so readialy hydrolysed as fats. Three common waxes, are as follow :

(a) Beeswax

(b) Lanoline

(c) Spermaceti.

Compound Lipids

These are the lipids which on hydrolysis yield other compounds, in addition to the alcohols and acids. They are of following types

Phospholipids

Phospholipids are soluble in both water and fat solvents, and, therefore serve an important role in binding both types of compounds together. They contain in addition to fatty acids and glycerol; phosphoric acid and a nitogenous base.

Three types of phospholipids have been indentified

(a) Lecithin. When placed in water, it swells up and readily forms an emulsion. Due to presence of acid and basic groups, it combines both with bases and acids. Thus these are important as a structural material in the cell membrane since by its hydrophilic group it maintains the continuity between the aqueous outside and the aqueous inside of the cell; yet fat soluble material dissolves in it and enters the cell because of its hydrophobic *group*.

(b) Cephalin. It is similar to lecithin. Cephalin and lecithin occur together in the tissue. *Howell* claims that the substance thrombokinase which initiates blood clotting is identical with cephalin. It is found specially in nerve tissues, *egg* yolk, etc.

(c) Sphingomyelin. These are derivatives of sphingosinol. [t is also found in nerve tissues, egg yolk etc.

(d). Plasmalogen. It is another type of phospholipid present in the brain, liver and muscle tissue.

Glycolipids

These are compounds of carbohydrates, fatty acids and sphingosinol which contain nitrogen, but no phosphoric group. They are also called as glycosphingosides. They are specially abundant in the brain.

Gangliosides

Found in nerve clls, spleen and red blood cells.

Lipoprotein

These are the complexes of lipids and proteins. Lipoprotein is present in cell membrane, egg yolk etc.

Sulpholipids

They are sulphuric esters of sphingosine cerebronic acid and galactose. Sulpholipids are observed in white matter of brain and to a lesser extent in liver, kidney, salivary gland, testes, etc.

Steroids

Steroids are not true fats. These are characterised by OH group thus are called sterols and are of highly importance for the body. Male and female sex hormones, adrenocortical hormones, vitamin D, cholesterol are major steroids. They are found in cell membranes and other cell structure containing lipids.

Carotenoids (Lipochromes)

These are red or orange cell pigments, soluble in organic solvents, but insoluble in water. This group *includes* the *Carotenes* in carrot and grass. *Xanthophyll in* leaves, vitamin A, egg yolk pigment. These are synthesized by the mitochondria of the cell.

NUCLEOTIDES AND NUCLEIC ACID

Nucleotides are involved in at least two major cellular functions: (1) they are monomeric units from which DNA and RNA polymers are constructed, and (2) they act as agents in certain energy-transferring reactions during metabolism. A mononucleotide is made up of one nitrogen-containing organic base, one pentose sugar, and one phosphate residue derived from phosphoric acid.

When there is no phosphate group, the sugar-base combination is called *a nucleoside*. For this reason, *nucleotides* (phosphate-sugar-base) are also called nucleoside phosphates. Nucleoside mono-, di-, and tri-phosphates contain one, two, or three phosphate groups respectively.

The nitrogenous bases commonly found in nucleic acids and their nucleotide building blocks are derivatives of purine and pyrimidine. The commonly occuring purines *adenine* and *guanine* are found in both DNA and RNA, as in the pyrimidine compound *cytosine*, the second kind of pyrimidine in DNA is *thymine*, while its demethylated form, *uracil*, occurs in RNA.

Table 1.3. Showing nomenclature of nucleic acids and their constituent units.

Base	*Nucleoside*	*Nucleotide*	*Nucleic Acid*
Purines			
Adenine (A)	Adenosine	Adenylic acid	RNA
Deoxyadenosine	Deoxyadenylic acid	DNA	
Guanine (G)	Guanosine	Guanylic acid	RNA
Deoxyguanosine	Deoxyadenylic acid	DNA	
Pyrimidines			
Cytosine (C)	Cytidine	Cytidylic acid	RNA
	Deoxycytidine	Deoxyadenylic acid	DNA
Thymine (T)	Thymidine	Thymidylic acid	DNA
Uracil (V).	Uridine	Uridylic acid	RNA

Since each kind of nucleic acid contains one unique pyrimidine, it is convenient to study synthesis and activity of DNA or RNA using isotopically-labelled precursors containing one or the other of these bases. Usually the nucleosides uridine or thymidine, or their nucleotide forms, are added to the biological system under study.

The only difference in the pentose sugars of nucleotides is the presence of a hydroxyl group at carbon atom 2 of D-ribose in RNA, but a hydrogen at carbon- 2 of 2-deoxy D-ribose in DNA monomers and polymers. This seemingly simple difference is partly responsible for profound differences in stabilities, pairing potential, and functions of DNA and RNA.

Polynucleotidcs of both DNA and RNA varieties are built from mononucleotides that are linked covalently via phosphodiester bridges between the 3' position of one unit and the 5' position of the next. Since there is no restriction on the vertical sequence of adjacent mononucleotides in either DNA or RNA, a considerable variety of molecules is possible even though only 4 kinds of nucleotide monomers (one for each of the four kinds of bases in the combination with sugar and phosphate) are used in polymer construction.

The theoretical variety is calculated as 4", where 4 is the number of different kinds of nucleotides, and n is the number of monomers in the polymer. For a molecule made of only 75 monomeric units, as in some of the smallest RNAs, there may be 4^{75} different arrangements of the constituent units.

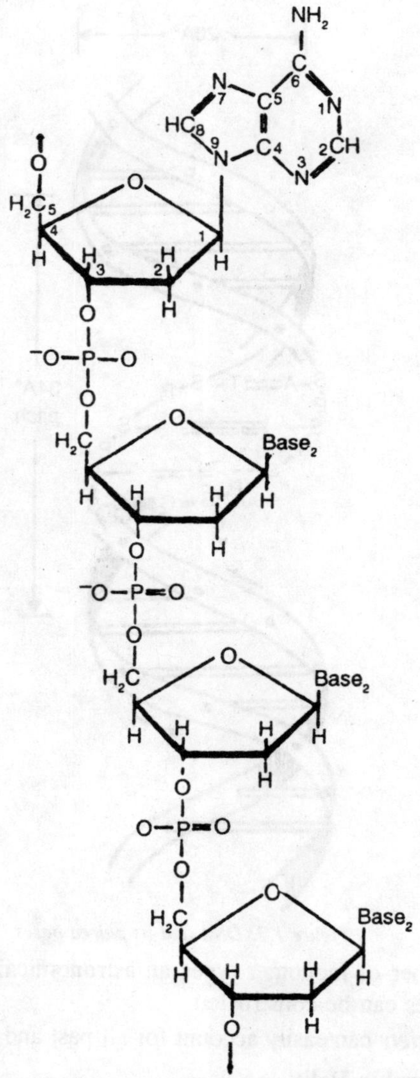

Figure 1.2 : The molecular organisation of a backbone of DNA showing phosphodiester and glycosidic bonds.

Each arrangement theoretically constitutes a molecule of different specificity.

Where the average gene may include about 500 nucleotides in a *DNA* sequence, 4500 different sequences are theoretically possible and, therefore that many different and specific genes. Despite the apparently

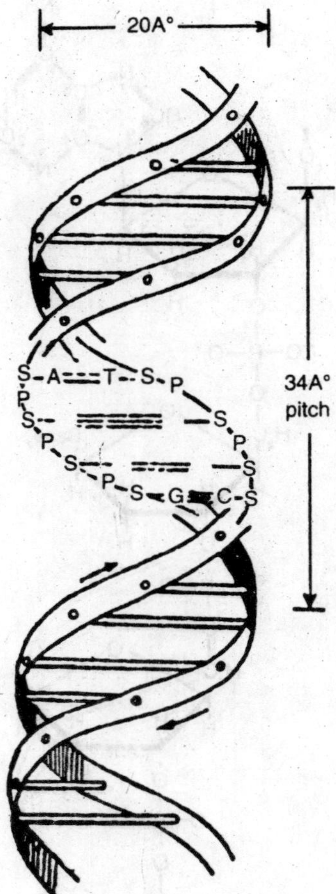

Figure 1.3: DNA and its paired bases.

meager number of monomer types an astronomically high number of possible genes can be constructed.

Such *variety* can easily account for all past and present life forms.

The DNA Double Helix

DNA molecules usually have regular helical configurations because most DNA molecules consist of two *complementary polynucleotide strands*. The two strands are held together by *hydrogen bonds* between complementary pairs of purifies and pyrimidines.

Adenine always binds with tbymine while guanine always binds with cytosine. This repeated hydrogen bonding within the double helix structure and bonding between virtually all the surface atoms in the

Ribose Sugar
Deoxribose Sugar

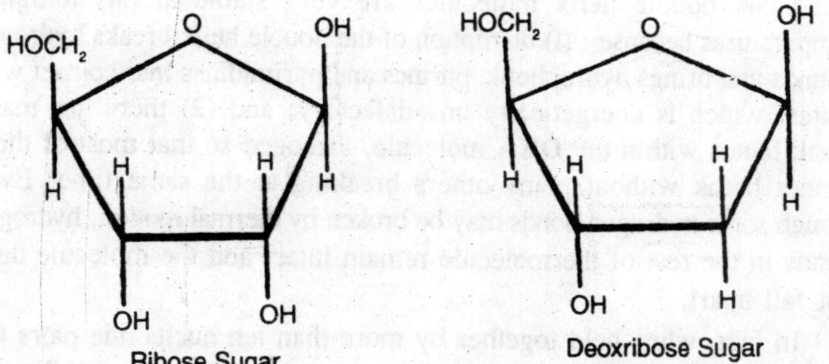

Figure 1.4: Pentose Sugar.

sugar and phosphate groups with water molecules serve to stabilize the structure.

Since the purine-pyrimidine pairs are found in the center of the molecule, their flat surfaces can stack on top of each other and thereby limit their contact with water. In rouble-helical molecules, a regular structure is possible because the complementary base pairs are exactly the same size.

Single polynucleotide chains could not have a regular backbone structure because pyrimidines are smaller than purines, which would cause the angle of helical rotation to vary with the sequence of bases.

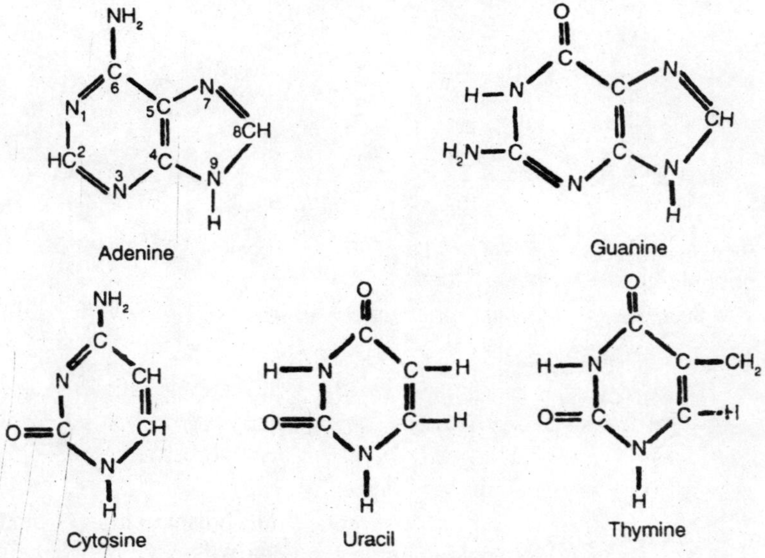

Adenine
Guanine
Cytosine
Uracil
Thymine

Figure 1.5: Various hetrogenous bases.

DNA double helix molecules are very stable at physiological temperatures because : (l) disruption of the double helix breaks hydrogen bounds and brings hydrophobic purines and pyrimidines into contact with water, which is energetically unsatisfactory; and (2) there are many weak bonds within the DNA molecule, arranged so that most of them cannot break without many others breaking at the same time. Even though some hydrogen bonds may be broken by thermal motion, hydrogen bonds in the rest of the molecule remain intact and the molecule does not fall apart.

In fact, when held together by more than ten nucleotide pairs the double helices are quite stable at room weak bonds is stability of molecular shape, in proteins as well as in nucleic acids. At abnormally high termperatures there is more frequent breakage of weak bond, which becomes less stable as temperatures rise above physiological levels. Once a significant number of weak bonds have been broken, a protein or nucleic acid molecule usually loses its original form and changes to an inactive or denatured form.

2

BIOCHEMISTRY OF PROKARYOTE

The smallest self-contained living entities governed by the genetic information of DNA are the bacteria. Because they are visible only when viewed under the microscope and do not contain *morphologically* distinct chromosomes that separate on mitotic spindles, their true nature was at first obscure.

The question long persisted as to whether bacteria, like all other cells, have a conventional hereditary apparatus based on genes or whether they represent some radically different form of life governed by interrelated *metabolic* pathways that utilize the laws of chemical kinetics to generate the cellular molecules needed for cell growth and division.

Today, the latter speculation appears at best a silly aberration by biologically naive *chemical kineticists*. However, only in 1943 did convincing evidence begin to appear supporting the existence in bacteria of discrete genes capable of spontaneous mutation.

Soon afterward, DNA was pinpointed as the carrier of genetic information and then localised in bacteria within distinct "*nucleoid*" bodies by the methods of cytochemistry. When the 1946 finding of genetic recombination in the bacterium *E. coli* opened up the possibility of systematic genetic crosses, research on bacteria acquired a momentum that only recently has begun to be challenged by efforts on other forms of life.

BACTERIAL NUCLEI

Although the cells of most bacteria are smaller than other types of cells, there is considerable size variation between the smallest and largest bacteria, whose cells approach the size of simple unicellular fungi (e.g., *yeast*). It is thus not size per se that makes a cell a bacterium.

Instead, it is the absence of a discrete nucleus that defines bacteria and their very close relatives, the blue-green algae (now often called *cyanobacteria*), as evolutionarily distinct forms of life. No nuclear membrane separates the DNA (chromosomes) of bacteria and blue-green algae from the cytoplasm in which protein synthesis occurs.

Organisms lacking nuclei are called *procaryotes*, while those organisms whose cells contain nuclei are called *eucaryotes (caryon* means "nucleus" in Greek). Even the smallest of eucaryotic cells differ quite fundamentally from procaryotic cells in the way their genetic information is organised as well as in their patterns of RNA and protein synthesis.

Bacteria accomplish these ends in more straightforward ways than eucaryotes, and it is unlikely that this simplicity is a matter of chance. While envolving to multiply more rapidly than any other cells, bacteria have become as simple as their nutritional requirements allow. It is this economy in their molecular components that made them the obvious organisms with which to initially establish the detailed chemical pathways through which genes control the life of the cell.

BACTERIAL GROWTH

There exists an enormous number of different types of bacteria, and they vary not only in size and shape but also in the nutritional conditions best suited for their growth and survival. Some bacteria, for example, are aerobes and grow only in the presence of oxygen; others are anaerobes and multiply only in the absence of oxygen; and still others, facultative anaerobes, can change their exact mixture of enzymes to allow growth in both environments.

While most bacteria derive their energy from breaking down externally derived food molecules, others have evolved photosynthetic pigments to let them use sunlight to make ATP. Independent of their nutritional specialisation, most bacteria grow free as single cells, separating from each other as soon as cell division Occurs.

In general, it has proved easy to grow them in the laboratory once their nutritional requirements have been worked out. Most importantly, in contrast to mammalian cells, which require a large variety of growth factors, many bacteria will grow well on a simple, well-defined diet, or medium. For example, the bacteria *Escherichia coil* (*E. coil*) will grow in an aqueous solution containing just glucose and several inorganic ions.

Table 2.1: Differences Between Procaryotic and Eucaryotic Cells.

Feature	*Procaryotes*	*Eucaryotes*
Genetic Organisation		
Nuclear membrane	Absent	Present
Number of different chromosomes	1	>1
Chromosomes with histones	Absent	Present
Nucleolus	Absent	Present
Genetic exchange	Plasmid-mediated, unidirectional	By gamete fusion
Cell Structures		
Endoplasmic reticulum	Absent	Present
Golgi apparatus	Absent	Present
Lysosomes	Absent	Present
Mitochondria	Absent	Present
Chloroplasts	Absent	Present in plants
Ribosome size	70S	80 S
Microtubules	Absent	Present
Cell wall with peptidoglycan	Present, except mycoplasma and archaebacteria	Absent
Some Functional Attributes		
Phagocytosis	Absent	Sometimes present
Pinocytosis	Absent	Sometimes present
Site of electron transport	Cell membrane	Organelle membranes
Cytoplasmic streaming	Absent	Present

Table 2.2: A Simple Synthetic Growth Medium for E. coil.

NH_4Cl	1.0 g
MgSO	0.13 g
KH_2PO_4	3.0 g
Na_2HPO_4	6.0 g
Glucose	4 0 g
Water	1000 ml

The growth of a specific bacterium is usually not dependent on the availability of a specific carbon source. Most bacteria are highly adaptable as to which organic molecules they can use as their carbon and energy sources. Glucose can be replaced by a number of other organic molecules, and the greater the variety of food molecules supplied, the faster a bacterium generally grows.

For example, if *E. coil* grows only on glucose, about 60 minutes are required at 37°C to double the cell mass. But if glucose is supplemented by the various amino acids and purine and perimidine bases, then only 20 minutes are necessary for the doubling of cell mass.

This shortening of *generation time* is due to the direct incorporation of the dietary components into proteins and nucleic acids, sparing the cell the task of carrying out the synthesis of their building blocks. There is a lower limit, however, to the time necessary for a cell generation; no matter how favorable the growth conditions, an *E. coil* cell is unable to divide more than once every 20 minutes.

GROWTH IN *E. coli*

The most intensively investigated and correspondingly best. known bacterium at both the biochemical and genetic levels is *E. coli*. Its prominence as the premier organism used to probe the essential features of life started in the early 1940s through its adoption by a group of young physicists, chemists, and geneticists.

These scientists believed that only by working on the simplest of biological systems would they ever be able to come to grips with the gene at its most fundamental. level. So they initiated research on *E. coli*, a common inhabitant of the human intestine. This organism possesses the very desirable properties of small size, lack of pathogenicity to most other organisms, and extreme ease of cultivation under laboratory conditions.

First extensively used to study the multiplication of its viruses, the bacteriophages (phages), *E. coil* became most interesting in its own right when it became the first bacterium shown to have discrete sexes that genetically recombine.

E. coli quickly became the obvious bacterium in which to search for mutations that blocked the synthesis of essential metabolites. Such mutations were relatively easy to find and nap, and they soon proved indispensable in the working out of the enzymatic steps through which bacterial molecules arc either made or broken down.

Today, virtually a thousand different *E. coil* genes have already been assigned chromosomal locations, and it would make no sense to start serious work on another bacterium if *E. coli* can be used to solve a particular problem. As we shall see in subsequent chapters, the combined methods of genetics and biochemistry are so powerful that it is undesirable to perform biochemical studies on an organism with which genetic analysis is not possible.

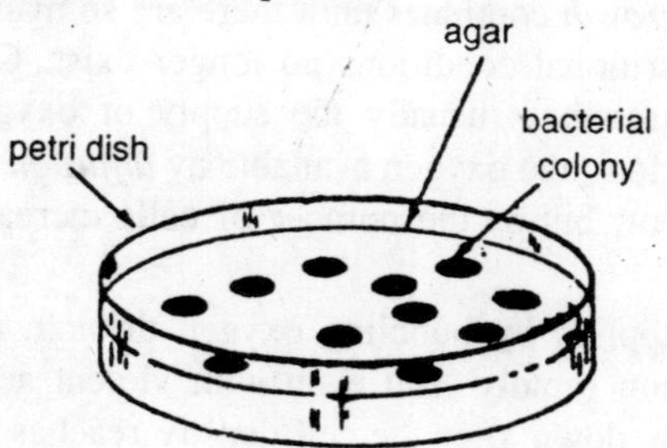

Figure 2.1 : The multiplication of single bacterial cells to form colonies. E. coli cells are usually not motile. Thus, when a cell divides on a solid surface, the two daughter cells and all their descendants tend to remain next to one another. After 24 hours at 37°C, each initial living cell gives rise to a solid mass of cells.

The average *E. coli* cell is rod shaped and about 2 μm in length and 1 μm in diameter. It grows by increasing in length followed by a fission process that generates two cells of equal length. Growth occurs best at temperatures around 37°C, suiting it for existence in the intestines of higher mammals, where it is frequently found as a harmless parasite.

It will, however, regularly grow and divide at temperatures as low as 20°C. Cell growth proceeds much more slowly at low temperatures; the generation time under otherwise optimal conditions is about 120 minutes at 20°C.

Cell number and size are often measured by observing the bacteria under the light microscope (and occasionally the electron microscope). Viewing a cell in this way, however, does not tell us whether the cell is alive or dead. We can determine whether a bacterium is alive or dead only by seeing whether or not it forms daughter cells. This observation is usually made by spreading a small number of cells on top of a. solid agar surface, which has been supplemented with the nutrients necessary for cell growth.

If a cell is alive, it will. grow to form two daughter cells, which in turn give rise to subsequent generations of daughter cells. The net result after 12 to 24 hours of incubation at 37°C is discrete masses, or *colonies*, of bacterial cells. Provided that the colonies do not overlap, each must have arisen from a single bacterial cell.

The growth of bacteria may also be followed in liquid, nutrient solutions. If a nutrient medium is inoculated with a small number of a rapidly dividing bacteria from a similar medium, the bacteria will continue dividing with a constant division time, doubling the number of bacteria each generation time. Thus, the number of bacteria increases in an exponential (logarithmic) fashion.

Exponential growth continues until there are so many cells that the initial optimal nutritional conditions no longer exist. One of the first factors to limit growth is usually the supply of oxygen. When the number of cells is low, the oxygen available by *diffusion* front the liquid surface is sufficient; but as the number of cells increases, additional oxygen is needed.

It is often supplied by bubbling oxygen through the solution or shaking the solution rapidly. But even with violent aeration, growth rates begin to slow down after the cell density reaches about 10^9 cells per milliliter, and a tendency develops for the cells being produced to be shorter.

Finally, at cell densities of about 5×10^9 cells per milliliter, cell growth is discontinued for nutritionally related reasons that are not yet clear. The term *growth curve* is frequently used to describe the increase of cell numbers as a function of time.

In most growing bacterial cultures, the exact division time of the cell varies, so that even if a culture has started from a single cell, after a few generations, cells can be found at various stages of the division cycle at any given moment.

Such growth is frequently called *unsynchronised growth*. Over the past ten years, tricks have been developed to isolate bacterial cells at the same stage of the cell cycle. These tricks can be used to obtain several generations of *synchronised cell growth*. Then, because of slightly unequal division times, the growth curve gradually acquires an unsynchronised appearance.

COMPLEXITY OF SMALL CELLS

Even cells as small as those of *E. coli* present great difficulties when we study them at the molecular level. At first sight, the problem of understanding the essential features of *E. coli* seems insurmountable, for on a chemical scale, even the smallest cells are fantastically large.

Although an *E. coli* cell is about five hundred times smaller than an average cell in a higher plant or animal (which has a diameter of approximately 10 μm), it nonetheless has a wet weight of approximately 10^{-12} gram (5×10^{11} daltons, where a dalton is the weight of one hydrogen atom).

Table 2.3: Inorganic Ions Found in a Bacterial Cell After Growth in a Glucose-Minimal Medium.

Ion	*Function*
K^+	Principal cation, cofactor for certain enzymes
NH_4^+	Principal form of inorganic nitrogen for assimilation
Mg^{2+}	Cofactor for a large number of enzymes
Ca^{2+}	Cofactor for certain enzymes
Fe^{2+}	Present in cytochromes and other enzymes
Mn^{2+}	Cofactor for several enzymes
Mo^{2+}	Present in several enzymes
Co^{2+}	Present in vitamin B_{12} and its coenzyme derivatives
Cu^{2+}	Present in several enzymes
Zn^{2+}	Present in several enzymes
Cl^-	Not required for many bacteria
SO_4^{2-}	Main source of sulfur in most media
PO_4^{3-}	Participant in many metabolic reactions

This number, which initially may seem very small, is immense on a chemist's scale, since it is 3×10^{10} times greater than the weight of a water molecule (MW = 18 daltons). Furthermore, this mass reflects the highly complex arrangement of a large number of different carbon-containing molecules.

There is also seemingly infinite variety in the chemical nature of the molecules contained in an *E. coli* cell. Fortunately, it is possible to distribute most of the large molecules into several well-defined classes possessing common arrangements of atoms. These classes are the proteins, lipids, carbohydrates, and nucleic acids.

Many molecules possess chemical groups common to several of

these categories, so the classification of such molecules is necessarily arbitrary. Also in the cell are many smaller organic molecules (such ns amino acids, purine and pyrimidine nucleotides, and various coensymes), very small inorganic molecules (e.g., O_2 and CO_2), and numerous electrically charged inorganic ions.

Among the very small molecules is water (H_2O), the most common molecule in all cells and a solvent for most biological molecules, through which diffusion from one cellular location to another can occur quickly.

At present, we can make only an approximate guess of the number of chemically different molecules within a single *E. coli* cell. Each year, many new molecules are discovered, so the known molecules clearly represent only a fraction of those we shall eventually know.

Already over 800 small molecules are implicated in its metabolism, and with a slightly large number of genes (and their proteins) already identified, we can say with confidence that *E. coli* must easily possess over 2000 different molecules.

This is surely an underestimate, and later we shall give reasons for suspecting that as many as 4500 different molecules might in fact be found within the average growing cell. Thus, we must immediately admit that the structure of a cell will never be understood in the same way that we understand water or glucose molecules. Not only will the three-dimensional structures of most cellular proteins remain unsolved, but their location within cells often remain imprecisely defined.

Table 2.4. The Various Molecules in a Bacterium Growing in Glucose-Minimal Medium.

Molecules	*Approximate Number of Kinds*
Amino acids and their precursors and derivatives	120
Nucleotides and their precursors and derivatives	100
Fatty acids and their precursors	50
Sugars, carbohydrates, and their precursors	250
Quinones, polyisoprenoids, porphyrins, vitamins, other coenzymes and prosthetic groups, and their precursors	300

It is thus not surprising that many chemists, after periods of enthusiasm for study "*life*", silently return to the world of pure chemistry. Increasingly, however, others are becoming lifelong converts to the fascination of biology once they realize (1) that all macromolecules are polymeric molecules built up from smaller monomers, (2) that there exist well-defined chains of successive chemical reactions in cells (metabolic pathways), and (3) that there is a limit to the number of proteins that can exist in a cell owing to the fàct that each cell contains a finite amount of DNA.

This number, in turn, leads to an estimate to the number of ensymes and hence small molecules that a cell can possess. Most likely, at least half the proteins in a cell are enzymes, each of which catalyzes a specific metabolic reaction. The approximate number of different types of small molecules could be estimated if we knew, on the average, how many specific enzymes are needed for the metabolism of the average small molecule. At present, it seems a good guess that the number lies between one and two.

THE ANATOMY OF *E. coli*

A direct way to appreciate the morphological simplicity of *E. coli* is through electron microscope examination of very thin sections cut from a rapidly growing cell. These sections reveal *E. coli* to have a simple, saclike structure in which an external envelope surrounds a membrane-free, dense granular cytoplasm (called the cytosol), which contains an apparently less dense fibrous DNA component.

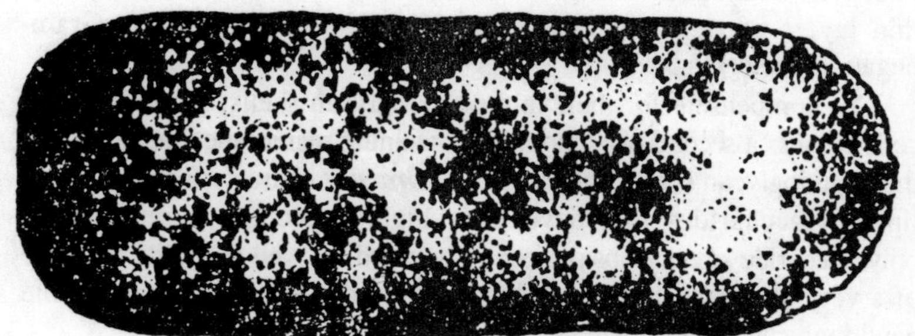

Figure 2.2: Electron micrograph of an E. coli B cell prepared substitution with no pretreatment.

The granular appearance of the cytosol reflects the very large number of ribosomes found in all growing bacterial cells. More ribosomes per unit dry mass are found in bacteria than in any other kind of cell.

This makes possible the high rate of protein synthesis needed to allow bacteria to double their number as rapidly as once every 20 minutes at 37°C. Very high concentrations of free enzymes also exist in bacteria, and the dry/wet ratio of bacteria can be as high as 30 percent, much greater than the 10 percent ratio characteristic of typical higher eucaryotic cells. The metabolic rate of bacteria is correspondingly much higher, with the amount of oxygen and substrates consumed per unit of time frequently 10 to 100 times greater than that consumed by eucaryotic cells.

BACTERIAL INDIVIDUALITY

The envelopes of bacteria are extremely complex compared to their tightly packed, relatively undifferentiated cytosol. The innermost layer of the envelope is always a phospholipid bilayer into which many functionally distinct proteins are inserted.

Surrounding this "*cytoplasmic facing membrane*" is a rigid shell of covalently linked carbohydrates and amino acids that has been given the name *peptidoglycan*. It is through their peptidoglycan component that bacteria obtain their structural integrity. If it is enzymatically removed (e.g., by *lysozyme*), the effectively naked bacteria assume spherical shapes. Depending on the specific bacterium, the peptidoglycan shell can be relatively thin (as in *E. coli*) or composed of many effective layers (as in bacteria belonging to the *Bacillus* groups).

Those bacteria that have thick peptidoglycan shells are very readily stained by the dyes crystal violet and iodine, first used by the Danish microbiologist Christian Gram. Such peptidoglycan-rich bacteria are known as Gram-positive bacteria, while those bacteria possessing only thin layers of rigid peptidoglycan are collectively known as Gram-negative bacteria.

To compensate for their thin peptidoglycan shells, Gram-negative cells possess a second phospholipid- containing membrane that surrounds the peptidoglycan shell. This *outer membrane is* characterised by unique lipopolysaccharides that have their highly specific (immunogenic) polysaccharide side groups projecting outward, as well as by the presence of a very abundant lipoprotein that anchors the outer membrane to the peptidoglycan shell.

Inserted into the outer membrane also are the matrix proteins, or *porins*, which from specific pores through which extracellular molecules must pass to reach the inner cytoplasmic membrane. Usually, such passages are passive processes governed by the laws of diffusion.

In other cases, movement through these pores is facilitated by at least 30 different *transport proteins*, or receptors, that make possible the entry of molecules too large for simple passage through the more common porin channels. How much transport proteins work remains to be established, with the possibility still open that they form specialised pores. Until recently, little was known about the structure of the "*periplasmic space*" between the inner and outer lipid-containing membranes. This region was known to contain three important classes of vital proteins: (1) hydrolytic enzymes, which initiate the degradation of food molecules; (2) specific binding proteins that help initiate the transport of certain food molecules across the inner cytoplasmic membrane; and (3) specific chemoreceptors used to measure the concentration of nutrients (or poisons) in the environment so that their respective cells can move toward (or away) from them.

Also, the periplasmic region was known to contain peptidoglycan, which was thought to exist in a thin, discrete layer near the outer membrane. Now, however, we believe that peptidoglycan acually fills the periplasmic space, forming a periplasmic gel that is highly fluid near the inner membrane and more compact near the outer membrane.

The gel is heavily hydrated, and periplasmic proteins can freely diffuse within it. Although it is generally thought that Gram-positive cells do not have an outer lipopolysaccharide layer, regular-appearing patterns suggestive of proteins have recently been seen on their outer surfaces. Whether these patterns are, in fact, proteins in a thin *lipopolysaccharide* bilayer remains to be worked out.

It is speculated that such a layer might not have been discovered earlier because of its relative sparcity compared to the thick peptidoglycan layer found in Grampositive cells. If this speculation proves correct, the difference between Gram-negative and Gram-positive cells will simply be a quantitative difference in the amount of pe.ptidoglycan in each

Unlike eucaryotic cells, where the aerobic generation of ATP is made possible by the passage of protons. through the membranes of mitochondria, the site of oxidative phosphorylation in bacteria is the inner, phospholipid-containing cytoplasmic membrane. Inserted within it are all the major proteins involved in the translocation of hydrogen ions, including the flavoproteins, quinones, iron-sulfur ("*FeS*") proteins, and cytochromes.

Their combined action leads to lower hydrogen ion concentrations in the cytosol compared to the exterior medium as well as differences in electrical charge across the inner membrane. Together they create the proton motive force, which provides the energy for a number of vital cellular processes that occur on the cytoplasmic membrane, including the active transport of food molecules and certain inorganic ions into the cytosol, the turning of flagella, and the generation of ATP from ADP by cytoplasmic membrane-bound ATP synthetase.

This latter enzyme is a complex molecular aggregate composed of at least ten different polypeptide chains, six of which combine to form knobs that protrude into the interior of the cell. The action of this membrane-based complex is reversible. The complex can either generate electrical charge differences across the cytoplasmic membrane (membrane potential) by hydrolysis of ATP to ADP (hence its alternative name, ATPase) or generate ATP from ADP and inorganic phosphate by the proton motive force. We know very little about the relative locations of the various proteins of the cytoplasmic membrane.

Studies using two-dimensional gel electrophoresis indicate that at least 200 different proteins are so located in the membrane; but this number is probably an underestimation of the true complexity. Clearly, we are just at the beginning of a real understanding of how membranes function at the molecular level.

Fortunately, for the first time, true crystals suitable for X-ray cystallographic examination have recently been obtained for several integral membrane proteins. Through the elucidation of their three-dimensional conformations, we may know within the next decade how molecules selectively move through membranes.

FLAGELLA AND PILL OF *E. coli*

E. coli, like many other bacteria, possesses two forms of extended surface appendages, flagella and pili. Though they are both long, linear bodies that arise from the cytoplasmic membrane, they are neither structurally nor functionally related. The bacterial *flagellum* (plural, *flagella*) is a very long, thin (200 Å diameter) organelle of locomotion, which is formed mainly by the helical aggregation of large numbers of a protein *building block called* flagellin. In reality, it is structurally quite complex.

It is composed of three parts : (1) the long filament made up of flagellin, which can extend 15 to 20 μm into the medium, *(2)* the *basal structure* that anchors the flagellum to the cell envelope, and (3) the hook, a short, curved structure that connects the filament to the basal structure.

Flagella serve as locomotory organelles by rotating the basal structures in either a clockwise or counterclockwise direction. In E. coli, counterclockwise rotation leads to the formation of a stable bundle of the several (about five) rotating flagella that propels the cell smoothly forward. In contrast, clockwise rotation, disperses the bundle and leads to a chaotic tumbling of the cell in many directions.

While the rotating filaments clearly function as propellers, the exact function of the hook is less clear. Like the filament, it is built up by the aggregation of a number of identical protein subunits. Much more complicated is the basal structure, which consists of at least ten different types of protein subunits that come together to form rings surrounding a thinner rod that is inserted into the cytoplasmic membrane.

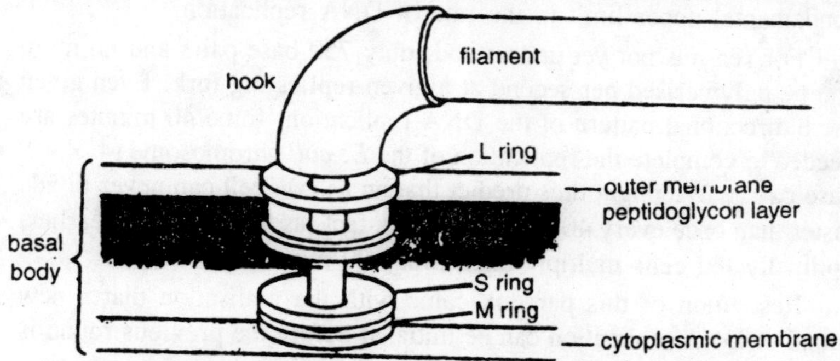

Figure 2.3: Diagram of the hook-basal body complex.

Exactly how the basal structure rotates remains to be worked out, though it is clear that the movement of hydrogen atoms across the cell membrane (the proton motive force) is involved. How specific attractants of repellents set into motion the reversals between clockwise and counterclockwise rotations has yet to be discovered.

Much simpler both in structure and in function are the *pilt* (singular, *pilus*). Composed of only single protein building blocks called pilin, they are adhesive organs that can extend 10 /hm or more into the medium. They function by attaching bacteria to other surfaces, often the glycoprotein components on the exterior of higher cells (such as the epithelial cells that line our intestines).

In addition to the more common type of pili present in large numbers on every bacterial cell, most male bacteria contain one to several morphologically similar pill called *sex pill*, which function to initiate contacts with female cells.

CONDENSATION OF BACTERIAL DNA

All the essential genes of *E. coli*. are present on a single chromosome, a single DNA molecule, which genetic experiments first revealed to be circular. Single cells, growing exponentially *trader favorable nutritional conditions*, *generally* contain two to four such molecules in the act of duplicating themselves, beginning at origins of replication present at one unique site on each chromosome. In such multiplying cells, the cell division process lags behind the DNA replication process, and only when cell growth has virtually ceased do, many *E. coli* cells contain only one chromosome. The reason why rapidly growing cells contain multiple copies of the same chromosome only recently became clear. It relates to the fact that there is a fundamental upper limit to the rate of DNA replication.

For reasons not yet understood, only 750 base pairs and no more can be polymerised per second at a given replicating fork. Even given the bidirectional pattern of the DNA replication, some 40 minutes are needed to complete the replication of the *E. coil* chromosome (4×10^6 base pairs). We might thus predict that an *E. coli* cell can never divide faster than once every 40 minutes. But in fact, as we mentioned earlier, optimally fed cells multiply once every 20 minutes.

Resolution of this paradox came with the realisation that a new round of DNA replication can be initiated before the previous round is completed. The replication of a given *E. coli* chromosome can thus occur during two successive cell cycles.

The length of a fully extended *E. coil* chromosome would be $4 \times 10^6 \times 3.4$ Å, or approximately 1 mm. However, it never even exists partially extended; electron microscopic examination reveals its compaction into irregularly shaped bodies, the *nucleoids*. Despite their similarity in name, nucleoids bear no relation to conventional nuclei, since they are not surrounded by any form of phospholipid-containing membrane.

Also distinguishing all procaryotic cells from their eucaryotic equivalents is the absence of conventional *histories*, the highly conserved class of basic proteins around which eucaryotic DNA coils to form the nucleosome particles that give to chromatin (the underlying DNA-protein complex of chromosomes) its granular (beaded) appearance.

Despite their lack of typical histones, however, gently prepared *E. coil* chromosomes initially display sections of beaded appearance that quickly transform into thinner fibers resembling pure DNA. So *E. coli* DNA must also, in part, be complexed with proteins that lead to its compaction.

In fact, two small histone like basic proteins have recently been found in *E. coil*, semitightly bound to the DNA. Their amounts, however, are insufficient to compact all the *E. coli* DNA into nucleosome-like particles. Instead, the most important factor that leads to the formation of nucleoid bodies may be enzymatically induced supercoiling, which of necessity leads to the compaction of the circular double helical DNA. Also still a mystery is how progeny bacterial chromosomes correctly partition themselves into their respective daughter cells. Evidence suggests that regions of each bacterial chromosome are associated with the inner, or cytoplasmic, mem brane of the bacterial envelope, with the occasional claim that the attached regions include the unique points where replication initiates (origins of replication). But even today, we have no good techniques to cleanly separate the cytoplasmic membrane from other cellular components.

TRANSCRIPTION AND TRANSLATION

At any given moment, the *E. coli* chromosome is being transcribed at the rate of 60 nucleotides per second (at 37°C) into some 400 to 800 unique mRNA chains, some 100 different tRNA molecules, and some 700 precursor molecules for rRNA. Each of these transcription events requires the participation of a single RNA polymerase molecule.

So at a given moment, some 1600 RNA polymerase molecules, representing about 1 percent of the total bacterial protein, are at work on a single *E. coli* chromosome. Even though only some 4 percent, of the total bacterial RNA is mRNA, its average *half-life of* perhaps no more than 12 minutes requires the participation of approximately one-half of a bacterial cell's RNA polymerase for its synthesis.

Very soon after mRNA chain *begins* to be made, its leading (5') end becomes attached to a free ribosome, which in turn commences to translate its message and in so doing to move toward the 3' end still in the process of extending itself along its DNA template. By the time a given mRNA chain is completely transcribed, it usually has attached to it a string of ribosomes successively carrying polypeptide chains or ever,ncreasing length.

With transcription occurring at approximately 60 nucleotides per second, the time required to make in average-size mRNA chain containing 2000 to 3000 nucleosides, approaches 1 minute, an interval not much shorter :han the average time (1.5 minutes) a given nucleotide remains incorporated within a given mRNA molecule. Thus, by the time the synthesis of many large mRNA molecules is completed, :hey may already have begun to be broken down.

A large fraction, if not a majority, of the ribosomes collected together on a single mRNA molecule (the assembly is called a *polyribosome* or *polysome*) may thus be physically attached to the bacterial chromosomes. Such linkages, however, do not trap significant numbers of ribosomes with DNA-containing nucleoide regions, since the extended lengths of average-size polysomes (10 to 15 ribosomes) approach the diameter of the *E. coli* cell itself.

The dilemma remains, however, as to how the nascent mRNA molecules with their attached strings of ribosomes separate from their DNA templates. In principle, this could occur either through rotation of the DNA or by the untwisting of RNA around DNA. Arguing against massiverotatory movements by the mRNA is the fact that many polyribosomes become securely fastened to the surrounding inner membrane so that their protein products can be built into the cell envelope.

But it is equally *unlikely* for RNA to separate away from DNA by a coupling of transcription to the rapid rotation events required for DNA replication. Not only does DNA replication occur at a rate some ten times faster than RNA synthesis, but also various genes along the *E. coil* chromosome are transcribed in different directions.

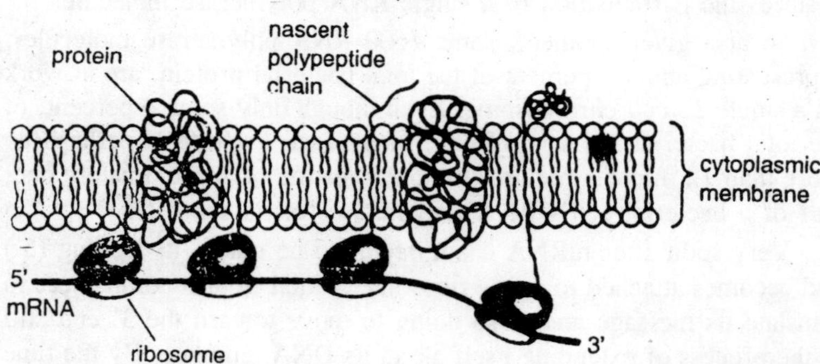

Figure 2.4: Synthesis of envelope proteins on polyribosomes attached to the cytoplasmic membrane. As translation proceeds, the nascent chain is transferred vectorially across the membrane bilayer. Its final destination may be the cytoolasmic membrane, the periplasmic space, or the outer membrane.

Perhaps those DNA segments being transcribed contain temporary singlestranded cuts along their backbone, which give them the capacity for the unimpeded retation needed to separate away from their RNA products. Slightly over half the dry mass of *E. coli* is protein.

Of this, some 20 percent is used to form the ribosomes, about 50 percent consists of the thousand or so enzymes found floating in the cytosol, and the remaining 30 percent is used to construct the cell envelope. Both the cytosolic proteins and the ribosomal proteins are made on free polyribosomes.

In contrast, the envelope proteins are made on polyribosomes attached to the inner (cytoplasmic) membrane of the envelope. Their peptide products are thus directly extended into the surrounding envelope.

Some of the completed protein ends up in the inner membrane and some in the periplasmic space. In addition, some of the protein first transferred to the inner membrane moves to the outer membrane, perhaps through tubular connections between the two membrane systems. Not only proteins, but all newly made phospholipids, appear first in the inner membrane before diffusing to the outer membrane.

METABOLIC PATHWAYS

We can see that all the molecules in a bacterial cell arise trom cellular transformation of food molecules when we grow *E. coli* on a simple, well-defined medium containing glucose. Under these conditions, glucose is the only organic source of carbon, and all the cellular carbon compounds (except for a few derived from carbon dioxide) result from enzyme-mediated chemical transformations that commence as the glucose molecules are broken down to smaller fragments.

These smaller molecules, in turn, are used to form the many diverse building blocks whose polymerisation creates the macromolecules from which major cellular organelles are assembled. The exact way in which all these transformations occur (collectively known as *metabolism*) is enormously complex, and most biochemists concern themselves with studying only a small fraction of the total interactions.

Fortunately, most of the major metabolic features of *E. coil* are common not only to all bacteria but also to all of life; therefore, what can be learned by focusing on *E. coli* is widely applicable. Figure shows some of the more important types of chemical reactions that interconnect the various small molecules of *E. coli*, starting with its intake of glucose.

Much information about this "*intermediary metabolism*" comes from experiments utilising radioactively labeled food molecules. For example, if we expose *E. coli* for several seconds (a "*pulse*") to ^{14}C-labeled glucose, the radioactive atoms can be detected almost immediately in molecules chemically similar to glucose, such as glucose-6-phosphate. Only later do labeled atoms find their way into amino acids and *nucleotides*. The amount of time before radioactivity appears in the various compounds corresponds roughly to the number of enzymatic reactions separating glucose from the various intermediate *metabolites*.

Such experiments are relatively easy to interpret because most of the small molecules within cells are represent in relatively small amounts. Only about 4 percent of the mass of *E. coli* is occupied by its collection of a thousand or so intermediary metabolites. Individual molecules thus have only fleeting existence before transformation into other small molecules or *polymerisation* into *macromolecules*.

The metabolic fates of the majority of molecules, however, are much more limited. An average molecules can be either broken down to compound *x* or used as an intermediate in the biosynthesis of comopund *y*; thus, each such metabolite is able to combine specifically with only two different enzymes.

Table 2.5. The Twelve Key Precursor Metabolites.

Glucose-6-phosphate
Fructose 6-phosphate
3-Phosphoglyceraldehyde
3-Phosphoglycerate
Phosphoenolpyruvate
Pyruvate
Acetyl-CoA
α-Ketoglutarate
Succinyl-CoA
Oxaloacetate
Pentose-5-phosphate
Erythrose-4-phosphate

All the biosynthetic pathways begin with one or another of a small group of molecules called *key precursor metabolites*. There exist just 12 of these precursor metabolites, from which some %5 building blocks (e.g., amino acids, purifies and pyrimidines, and fatty acids) and coenzyme products are derived by specific sets of enzymatic reactions.

These pathways, however, are not universally present in all bacteria. Many bacteria lack one or more specific biosynthetic routes and correspondingly require the respective end product(s) to be externally supplied for growth. Not surprisingly, it is those bacteria normally found in environments rich in organic material that have lost one or more specific biosynthetic pathways.

When such a pathway is present, however, it is almost always composed of the same set of enzymatic reactions, which either begin directly with one of the 12 key precursor metabolites of branch off from an intermediate or end product of another pathway.

DEGRADATIVE PATHWAYS

When *E. coli* is growing with glucose as its *sole* carbon source, all its amino acids must be synthesized from metabolites derived from glucose. There is a distinct biosynthetic pathway for each of the 20 amino acids. But *E. coli* can also grow in the absence of sugar, using any of the 20 amino acids as a sole carbon source. This means that there must also exist 20 pathways of amino acid degradation by which the carbon and nitrogen atoms of the amino acids are usually freed to form key metabolite compounds such as α-ketoglutarate.

These compounds can then be used in the synthesis of other amino acids. Degradative pathways also exist for the various lipids, the purine and pyrimidine nucleotides, many pentose and hexose sugars, and so on.

That distinct degradative pathways are almost never simply the reverse of biosynthetic pathways is to be expected. As stated already, most biosynthetic reactions require energy and often involve the breakdown of ATP, whereas degradative reactions by their very function must eventually generate ATP in addition to supplying carbon and nitrogen skeletons.

DIVERSITY OF FUELING REACTIONS

The degradative reactions that produce the 12 key precursor metabolites, ATP, and the needed reducing power (NADH) are called *fueling reactions*. While many fueling reactions are common to all bacteria, others are unique to particular species. Those fueling reactions common to all cells, called the *central pathways*, are involved in the derivation of the key metabolic precursors from each other.

They can function linearly, as in glycolysis, to replace key metabolites drained off for biosynthesis, or they can function cyclically, as in the Krebs cycle, to produce carbon dioxide and water as well as to generate ATP and NADH. Thus, there must be very great flexibility in the way the central pathways operate.

The real diversity of the fueling reactions is expressed by the peripheral pathways used when bacteria grow on compounds that are not intermediates in the central pathways. The enormous variety of potentially valuable sources of carbon, nitrogen, energy, and reducing power has led to the evolution of the approximately 4000 known forms of bacteria that can utilise almost every carbon compound naturally existing on Earth with the exception of diamonds and coal.

Even within a single type of bacterium, the number of alternative fueling reactions that can be called into action, depending on the nutritional environment, is enormous. *E. coli* probably possesses at least 75 different peripheral pathways made up of at least 200 to 300 different enzymatic reactions. These numbers must be even larger in many other bacteria, particularly the pseudomonads, which can grow on a much larger, more varied collection of inorganic molecules than can *E. coli*.

PERIPHERAL PATHWAYS

At any given time under constant nutritional conditions, a bacterial cell may possess the enzymes for only a single peripheral pathway. Even when several different sugars are available as food sources, an *E. coli* cell tends to have only the enzymes for the pathway that generates the most energy. For example, in the presence of both glucose and galactose, *E. coli* only possesses the enzymes needed to break down glucose.

Only when the glucose supply is exhausted does it make the enzymes needed to metabolise galactose. The set of enzymes that forms a peripheral pathway is thus present or absent as a unit. This makes obvious biological sense, for it would be wasteful to possess only half a pathway and thereby produce intermediate compounds that cannot be used. Likewise, it would make no sense to possess the enzymes needed to metabolise food molecules that are not present in the environment.

The genes coding for enzymes of peripheral pathways constitute a significant fraction of the *E. coli* genome, and if they functioned in the absence of need, a significant fraction of the cell's protein-synthesizing capacity would be tied up to no avail.

REGULATION OF PROTEIN FUNCTION

The catalytic activity of many proteins is affected by their ability to bind to specific small molecules. Because of this property, the activity of enzymes may be blocked when they are not needed. Consider, for example, what happens when an *E. coli* cell growing on minimal glucose medium is suddenly supplied with the amino acid isoleucine.

Immediately, the synthesis of the mRNA molecules that code for the specific enzymes utilised in isoleucine biosynthesis ceases. Without a further control mechanism, preexisting ensymes will cause continued *isoleucine* production, now unnecessary because of the extracellular supply. Wasteful synthesis, however, almost never occurs, *because high* levels of isoleucine block the activity of the enzyme involved in the first step of its biosynthesis from threonine.

This inhibition is due to the binding of isoleucine to the enzyme threonine deaminase. Thus bound, the enzyme is unable to convert threonine to a-ketobutyrate. Because the association between the enzyme and isoleucine is weak and reversible, relatively high isoleucine concentrations must exist before most of the enzyme molecules are inactivated.

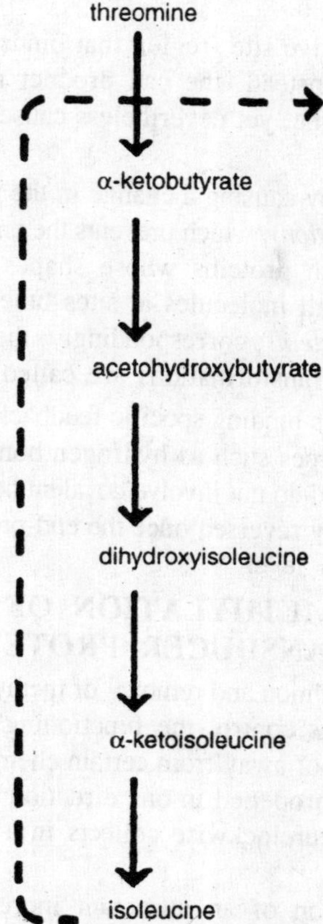

Figure 2.5: The pathway of isoleucine biosynthesis starting from threonine. The dotted colored line shows that isoleucine inhibits the enzyme (threonine deaminase) that transforms threonine into α-ketobutyrate.

This very specific inhibition is called *feedback inhibition* or end-product inhibition because accumulation of a product prevents its further formation. Usually, only the first step in a metabolic chain is blocked. With the first reaction blocked, there is no accumulation of unwanted intermediates; so inhibition of the remaining enzymes would serve no purpose.

The final enzymatic step in the synthesis of a feedback inhibitor is often separated by several intermediate metabolic steps from the substrate (or product) of the enzyme involved in the first step of its biosynthesis. Figure gives another example of this situation. The structure of the inhibitor may thus only loosely resemble that of the substrate of the inhibited enzyme, so that we would not expect an end-product inhibitor to combine with the active site (region that binds the substrate) of the enzyme it inactivates Instead, the cad product reversibly binds to a second site on the enzyme, yet nevertheless causes the enzyme activity to be blocked.

The inhibitor acts by causing a change in the precise enzyme shape (an *allosteric transformation)*, which prevents the enzyme from combining with its substrate. Such proteins whose shapes are changed by the binding of specific small molecules at sites other than the active site are called *allosteric proteins*; correspondingly, the small molecules that bring about allosteric transformations are called *allosteric effectors*.

The chemical forces binding specific feedback inhibitors to proteins are weak secondary forces such as hydrogen bonds, salt linkages, and van der Waals forces and do not involve covalent bonds. Hence, feedback inhibition can be quickly reversed once the end-product concentration is again reduced to a low level.

METHYLATION OF TRANSDUCER PROTEIN

It is through the addition and removal of methyl groups that chemical attractants or repellents control the functioning of bacterial flagella. This movement toward or away from certain chemicals is called *chemotaxis*. *E. coli* cells are propelled in one direction when a set of flagella rotating together counterclockwise collects in a bundle that acts as a propeller.

If the concentration of an attractant increases, this rotation is prolonged, so that a cell tends to "*run*" ("*swim*") toward the attractant. If a repellent is encountered, on the other hand, the direction of rotation is reversed to clockwise, destroying the efficiency of the propeller and causing the cell to "*tumble*" for a while. When tumbling stops, the cell is likely to move off in a different direction and so escape the repellent. When no chemical is present, the cell is in a neutral state in which it alternately runs and tumbles.

For many attractants and repellents, the signal to swim or to tumble is transmitted from the outside into the cell by *transducer proteins* that span the cell's cytoplasmic membrane. The protein's NH,-terminal portion extending outside the cell either binds the attractant (or repellent) directly or is bound by other periplasmic proteins that detect the chemical.

In either case, when the transducer protein is contacted from the outside, it undergoes some change in its intracellular (COOH terminal) portion that ultimately signals the flagellar motor to go either counter clockwise (in response to an attractant) or clockwise (in response to a repellent). What this change might be is completely unknown; one guess, however, is that an attractant or repellent stimulates an enzymatic activity in the intracellular domain of the transducer to synthesize a signal molecule that diffuses to the flagellar motor and switches it from one direction to the other.

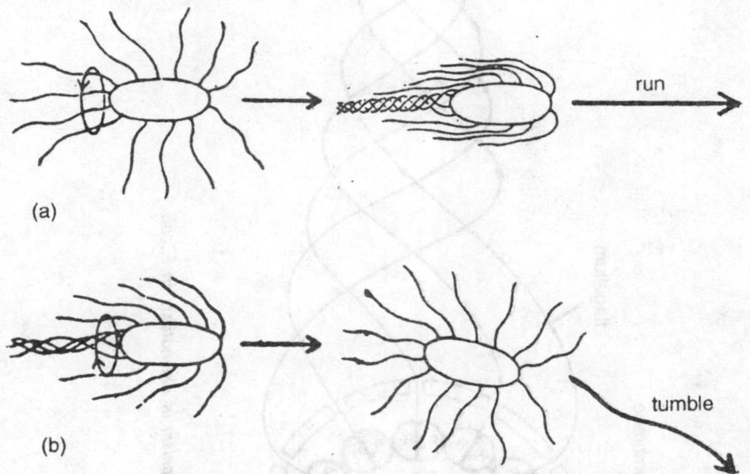

Figure 2.6: Mechanism of movement during chemotaxis in E. coli. (a) An attractant causes a counterclockwise rotation of the flagella, which form a bundle to propel the cell forward. (b) A repellent causes the bundle to rotate clockwise, the bundle falls apart, and the bacterium tumbles.

How the cell actually behaves in a complex environment depends on the combined effects of all signals sent to its motor by all transducing proteins (and certain other signaling systems) that may be sensing a variety of chemicals.

A further property of the transducer protein is *adaptation*. If the new concentration of attractant or repellent persists, the transducer soon adapts to this new condition and no longer transmits a signal. This means that the transducer responds to change and not to a given concentration of an attractant or repellent.

Adaptation is caused by modification of a transducer through methylation and demethylation of about four glutamyl groups on its cytoplasmic side. Methylation favours the state in which a transducer signals clockwise rotation (tumbling), whereas demethylation promotes counter-clockwise rotation (running).

The modification that occurs opposes the last chemotactic signal, but takes about a minute to occur. Thus, exposure to an attractant

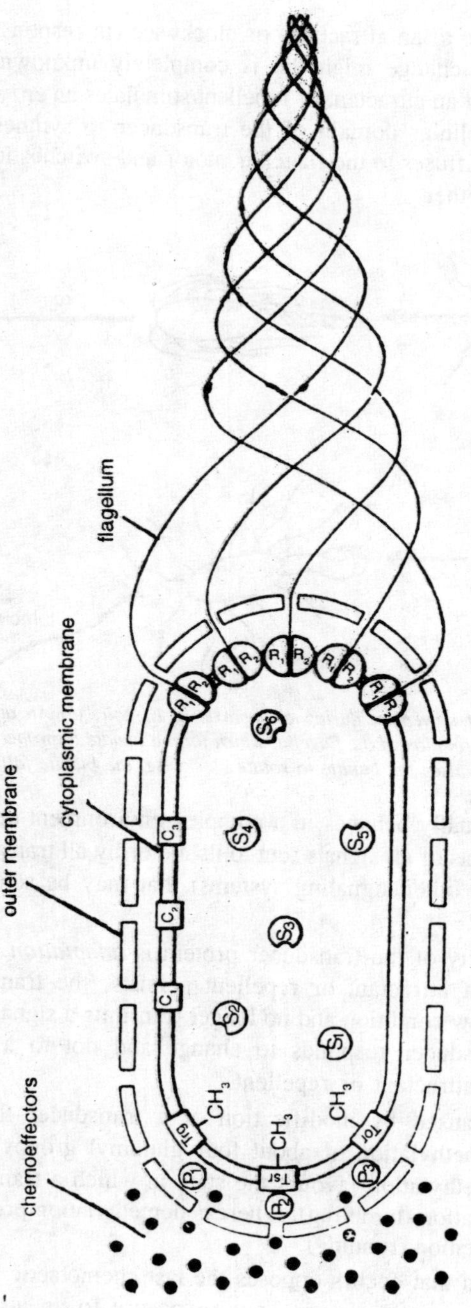

Figure 2.7: The mechanism of chemotaxis in E.coli.

eventually leads to greater methylation, allowing more tumbling again. And exposure to a repellent leads to demethylation, so that tumbling once again alternates with periods of swimming. Probably the same conformational change of a transducer protein that produces a signal to the flagellar motor also allows it to bind either the methylation or demethylation enzyme, so that the initial effect of the attractant or repellent is inevitably neutralised.

3

Chemical Energy Fuel

Carbohydrates constitute by far the greatest proportion of organic material on the face of the earth and the most abundant carbohydrate is cellulose which forms the main supporting structure of plants. They constitute important materials for the necessities of life such as food and clothing, housing and health. As reserve food material for plants, the starches are stored in grains, tubers and roots and form the staple food and main energy sources of the poorer populations of the world.

The sugars are found in the necter of flowers and in fruits and milk. Among the industrial processes based on carbohydrates may be mentioned textiles, paper, plastics, explosives, *fermentation* industries and alcohol. The vitamin, ascorbic acid, is a carbohydrate deivative and some of the plasma expanders belong to the class of carbohydrates. *Glycosides* of the digitalis-strophanthus group are used in medicine and it is recognised that the *invasiveness* and *antigenic nature* of some bacteria is due to their *polysaccharide* envelope.

A great many of the wondrous colours of the plant kingdom are due to the glycoside pigments of the anthocyanin and flavone groups. From the theoretical standpoint, the genesis of carbohydrates in plants through the mechanism of *photosynthesis*, the complicated steps in the enzymatic breakdown of the simple sugar glucose to carbon dioxide and water, and the *phenomena* of muscular contraction and bacterial fermentation have fascinated many generations of biochemists. Inside the animal cell, the storage form of energy is present in the form of glycogen.

The *ubiquitous nucleic acids*, deoxyribonucleic acid in the nucleus and ribonucleic acid in the cytoplasm and nucleolus, derive their name from the carbohydrate present in them. The nourishment of the cell come from the circulating glucose of the extracellular environment and the activities of the cell may be profoundly altered by the *mucopolysaccharides* of the surrounding extracellular tissue or the bloodborne glycoprotein hormones of the anterior pituitary. *Mucopolysaccharides* are also biological lubricants in the form of saliva, *mucus* and *synovial fluid*.

The term carbohydrate originated in the belief the naturally occurring compounds of this class could be represented by a general formula $C_x(H_2O)$. Though *formaldehyde*, *acetic acid*, and *lactic acid* fulfil this formula requirement, they are not carbohydrates and some of the methylpentoses and other "*deoxy*" sugars which belong to the carbohydrate class would be excluded by this requirement. The term carbohydrate is retained, therefore, as a matter of convenience rather than of precise definition.

The term carbohydrate now encompasses a variety of substances which include *polyhydroxy aldebydes* and *ketones*, *alcohols*, *acids*, *amino sugars* and their derivatives as well as the numerous products formed by the condensation of these different types of compounds with each other by means of glycosidic linkages to form *oligosaccbarides* and polysaccharides. Polyol (*glycerol* and *ribitol*) phosphate polymers like teichoic acids are also included in this class of compounds. The elucidation of their chemistry has been a saga of great effort in organic chemistry and even today presents an area of challenge and great biological significance.

The structures of eight aldohexoses were carefully elucidated of which only three occur naturally while the rest were synthetic. Now over two hundred different monosaccharides are known to be present in nature. Many amino sugars, branched chain sugars and a wide variety of N-acetyl N-glycolyl *neuraminic acids* (*sialic acids*) have been isolated. Several antibiotics like streptomycin, puromycin, neomycin and aminoglycoside antibiotics like *kanamycin* certain unusual types of sugar residues. The glycoprotein hormones include *gonadotropic hormones*, *thyroglobulin* and *thyrotropin*. Anti-freeze glycoproteins in aquatic animals in cold environments have aroused attention.

The characteristic bacterial cell wall peptidoglycans linked to peptides and their susceptibility to *penicillin* have been extensively studied. Lipid-linked sugar derivatives linked by phosphate bridges to long chain unsaturated lipids (*polyprenols*) offer clues to the nature of binding in glycolipid molecules. Enzyme defects and lysosomal membrane alterations present in mucopolysaccharide disorders like Hurler's syndrome have provoked attempts to induct enzymes to correct these conditions. The non-enzymatic glucosylation of hemoglobin, serum proteins, and lens proteins with an increased incidencc of cateract in diabetic subjects suggests that increased formation of glycoproteins could produce changes in the basement membrane of the smaller blood vessels and give rise to complications.

Purified polysaccharides like those isolated from pneumococcus type III have been shown to be antigenic. Studies using specific *glycosidases* and *phytohaem agglutinins* (*lectins*) with an affinity for certain sugars have revealed that blood group specificities are related to the presence of particular sugar residues on the glycoproteins associated with them. The understanding of the properties of cell membranes depends on the knowledge regarding the molecular structure of the glycoproteins and glycolipids and particularly the disposition of the sugar residues present on their surface.

Sugar residues are attached in glycoproteins by *glocosylamine* link to the amide nitrogen of specific asparagine residues in the polypeptide chain or through the hydroxyl groups of serine, threonine or hydroxylysine. Sialylation of proteins has far-reaching effects on their biological properties. Studies in carbohydrate chemistry, now facilitated by the use of specific enzymes, chromatographic separation procedures, spectrometry of different kinds and X-ray diffraction, may illuminate the problems involved in cell regulation and immunity in general.

Carbohydrates may be broadly classified into three main groups

A. Mouosaccharides, e.g., ribose, glucose, glactose, sedoheptulose.

B. Oligosaccharides with 2-10 monosaccharide units e.g., sucrose, lactose.

C. Polysaccharides, e.g., glycogen, starch, cellulose, insulin.

Owing to their sweet taste and cystalline character the mono- di- and trisaccharides are generally grouped together under the name of *sugars*. The ploysaccharides are tasteless, amorphous and sparingly soluble substances which on hydrolysis yield monosaccharides.

CLASSIFICATION

Carbohydrates, or *saccharides*, are *polyhydroxyaldehydes* or *polyhydroxyketones* and their derivative. The simplest of these, the *monosaccha-*

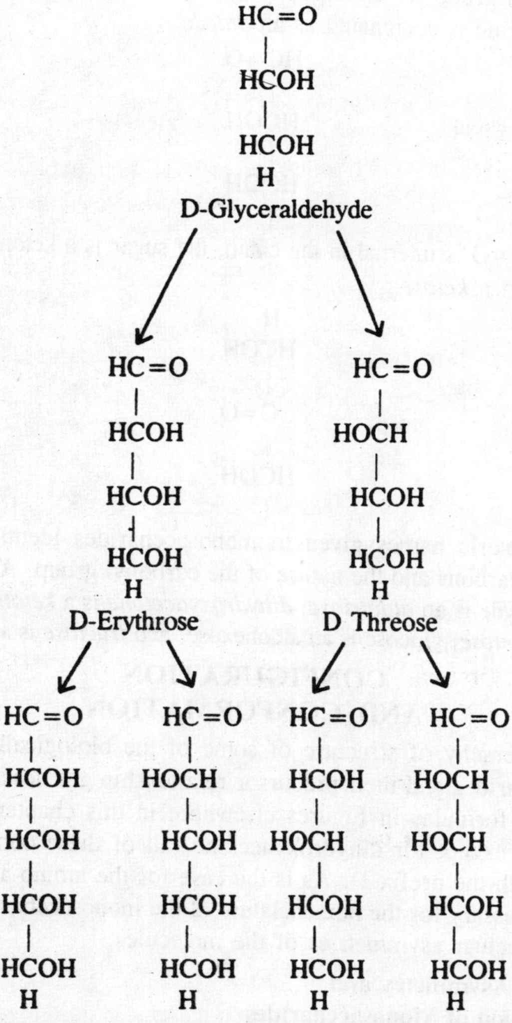

rides, may be considered to be hydrates of carbon —$(CH_2O)_n$, for which n is 3 to 8. *Oligosaccharides* are *carbohydrates* comprised of a multiple (2 to 10) of monosaccharides joined by acetal or ketal linkages. Polymers of saccharides exceeding 10 in number are referred to as *polpsaccharides*; the chains of monosaccharide units in these molecules may be linear or branched.

The carbon chain of a monosaccharide is usually unbranched. Each carbon atom, except one, carries a —OH group ; the remaining carbon is a carbonyl group. If the carbonyl group is at a chain terminus, the monosaccharide is designated as an *aldose*:

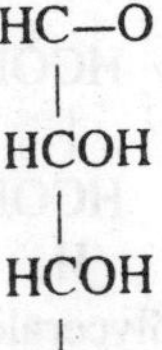

If the C=O is internal in the chain, the sugar is a ketone derivative and is called a *ketose*:

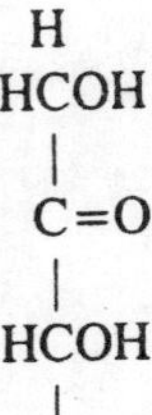

The generic names given to monosaccharides identify both the number of carbons and the nature of the carbonyl group. As examples, glyceraldehyde is an *aldotriose*; *dihydroxyacetone* is a *ketotriose*; ribose is an *aldopentose*; glucose is an aldohexose; and *fructose* is a *ketohexose*.

CONFIGURATION AND CONFORMATION

The hierachy of structure of some of the biologically important monosaccharides and their precursor relationship are described by the open-chain formulas in Figures elsewhere in this chapter. It will be noted that, except for dihydroxyacetone, all of these compounds are labelled with the prefix D. As is the case for the amino acids, such a label is necessary for the nomenclature of the monosaccharides because of the molecluar asymmetries of the molecules.

Molecular Asymmetry and Configuration of Monosaccharides

Except for *dihydroxyacetone*, all of the *monosaccharides* described in Figures are chiral molecules. All contain one or more *asymmetric carbon atoms* (i.e., a carbon with four different substituents) and can therefore exist in right and left-handed forms. Such isomers are nonsuperimposable mirror images of each other and are called *stereoisomers* or *enantiomers*. Stereoisornerism is evident by the isomers optical activity, the ability to rotate the plane of plane-polarised light. Enantiomers are identical in all physical and chemical properties, except for their action on polarised light. Stereochemical properties influence the molecular structures and metabolic behaviour not only of carbohydrates but also of amino acids, peptides and proteins.

D-Ribose		D-Arabinose		D-Xylose		D-Lyxose	
↙	↘	↙	↘	↙	↘	↙	↘
HC=O	HC=O	HC=O	HC=O	HC=O	HC=O	HC=O	HC=O
HCOH	HOCH	HCOH	HCOH	HOCH	HOCH	HCOH	HOCH
HCOH	HCOH	HOCH	HOCH	HCOH	HCOH	HOCH	HOCH
HCOH	HCOH	HCOH	HOCH	HOCH	HOCH	HOCH	HOCH
HCOH	HCOH	HCOH	HCOH	HCOH	HCOH	HCOH	HCOH
HCOH	HCOH	HCOH	HCOH	HCOH	HCOH	HCOH	HCOH
H	H	H	H	H	H	H	H
D-Altrose	**D-Altrose**	**D-Glucose**	**D-Monnose**	**D-Gulose**	**D-Idose**	**D-Galactose**	**D-Talose**

Figure 3.1: D-Aldoses.

Glyeeraldehyde is the smallest sugar with an asymemetric carbon atom. To indicate the tetrahedral nature of the carbon atom and the spatial consequences of four different subtituents on one carbon atom, consider glyceraldehyde to be a *tetrahedron* as gwn in Figure. The molecular asymmetry of this molecule is apparent rom the perspective and projectionformulas shown in Figure elsewhere in this chapter. Because of the tetrahedral character of the carbon atom, the four different substituent groups on the center corbon of glyceraldehyde can have two different spatial arra-ngements.

These two structural configurations represent the enantiomers of glyceraldehyde since the two forms are non-superimposable. To avoid the three-dimentional rojections in writing formulas as stereoisomers, it is generally lore convenient to use two-dimentional representations, called *sher projections*. The formulas of the two glyceraldehydes written in this style are shown in Figure. In using Fisher proctions, the formulas cannot be removed from the plane of the aper when determining if two molecules are superimposable.

In describing the *asymmetry* of the glyceraldehyde molecule we are concerned with the *configuration* of the substituents about the center carbon. As seen in Figure two configurations are possible when a molecule has one asymmetric carbon. These two forms are designated as D- and L-. The D- and *Lglycerıldehydes* have been made the reference

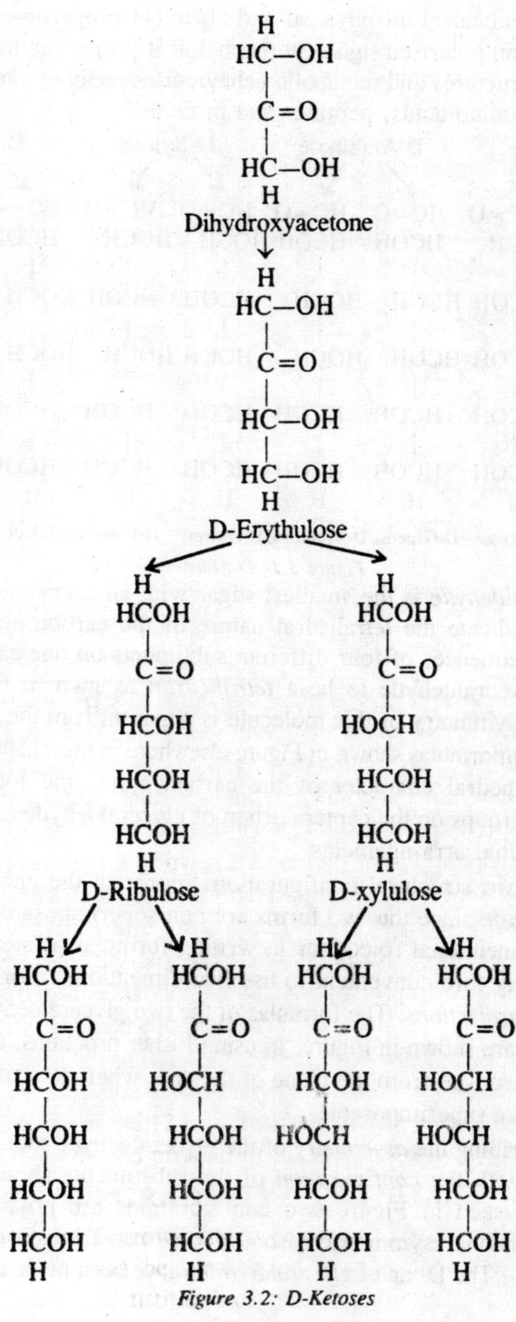

Figure 3.2: D-Ketoses

or parent compounds for designating the absolute configurations of all stereoisomers. By convention the projection formulas for the D- and L-glyceraldehydes arearawn as shown in Figure elsewhere in this chapter.

All compounds whose stereochemical configurations are related to D-glyceraldehyde are labelled as D forms and all those related to L-glyceraldehyde are designated as L-compounds, irrespective of the direction of rotation of plane polarised light. Optical activities are designated as, d, or (+), or dextrorotatory, and 1, or (-), or *levorotatory*. In the case of the glyceraldehydes, the directions of rotation are such that the designations of optical activity and configuration are similar [i.e., *D* (*d* or +)-glyceraldehyde and L (1 or—)-glyceraldehyde].

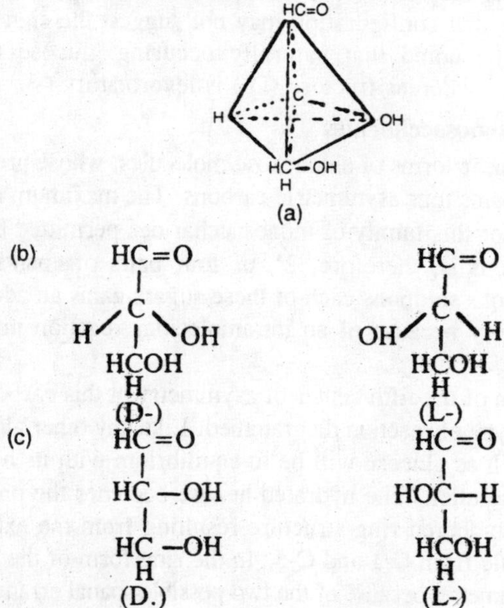

Figure 3.3 : Stereochemistry of glyceraldehyde.

L-Lactate, a common human metabolite, and L-alanine are *dextrorotatory*. Therefore, these compounds are designated as L (+)-lactate and L (-)-alanine. Obviously, *stereochemical* complexity increases in compounds containing more than one asymmetric carbon atom. For n number of asymmetric centers the number of *stereoisomers* will be 2", or n pairs of enantiomers. For example, the aldohexoses have eight pairs of enantiomers or optical isomers, as shown in Figure. The sugars shown here which are not (+)/(-) are called *diastereoisomers*; the eight aldohexoses given in Figure are *diastereoisomers*. If two diastereoisomers differ only in the configuration around one carbon atom [e.g., carbon 2 (C-2) of D-glucose and D-mannose, or C-4 of D-glucose and D-galactose], they are referred to as *epimers*. How is the configuration of such isomers designated ? By convention, the asymmetric carbon most remote from the carbonyl carbon (i.e., the C=O) is the reference carbon. Thus, a compound will be designated D- or L-, depending on the configuration of this carbon. Note, for example, that the G-5's of all of the hexoses in figure elsewhere in this chapter have the same configuration as the center carbon in D-glyceraldehyde.

Therefore, all of these sugars are of the D series. The same relationship obtains for the tetroses and pentoses. To underscore the point made earlier that configuration may not suggest the sign optical activity, it may be noted that naturally occuring glucose (D-) is dextrorotatory (+), whereas fructose (D-) is levorotatory (-).

Ring Forms of Monosaccharides

Each of the linear forms of aldohexose molecules, whose projection formulas are, contains four asymmetric carbons. The maximum number of stereoisomers for this family of monosaccharides permitted by their linear configurations is, therefore, 2^4, or four pairs of enantiomers. However, in aqueous solutions each of these sugars gains an additional center of asymmetry because of an intramolecular reaction involving the carbonyl group.

The formation of the fifth center of asymmetry at this carbon atom can occur by the type of reaction diagrammed. Like any other aldehyde, an aldohexose such as glucose will be in equilibrium with its hydrated form in aqueous solution. The hydrated hexose becomes the precursor of a stable, six-membered ring structure resulting from the extraction of a water molecule from C-1 and C-5. In the ring form of the aldose. C-1 becomes asymmetric because of the two possible spatial arrangements of its -OH group.

Employing plane projection formulas, the isomer in which the -OH is in the same side as the ring oxygen is designated as the *(z- configuration*. If the -OH group is on the side opposite to the ring oxygen the isomer is referred to as a *β form*. The sixmembered ring structures of this type are related to the heterocyclic compound pyran and, hence, the sugars are labelled as pyranoses.

The two structures are α-glucopyranose and β-D-glucopyranose. With the additional center of asymmetry at C-1, each D- and L-aldohexose can have an α or a β form, allowing at a total of 2^5 or 32 aldohexoses. Monosaccharides that differ only with respect to the configuration of their carbonyl carbon atoms (i.e., α or β) are known as *anomers*. To appreciate the elementary chemistry of pyranose formation and its implications for our understanding of carbohydrate conformation, consider again the analogous and simpler reaction between an alcohol and an aldehyde or a ketone. Under conditions in which the aldehyde reacts with the equivalent of one alcohol to form a hemiacetal, the erstwhile aldehyde carbon will carry four different substituents and, therefore, becomes asymmetric.

HC—O
HCOH
HOCH
HCOH
HCOH
HCOH
H

HOH

α	β
OH	OH
HC—OH	HO—CH
HCOH	HCOH
HOCH	HOCH
HCOH	HCOH
HCOH	HCOH
HCOH	HCOH
H	H

hydrated D-Glucose

HOH HOH

α	β
HC—OH	HOCH
HCOH	HCOH
HOCH	HOCH
HCOH	HCOH
HC	HC
HCOH	HCOH
H	H

O O

Figure 3.4: α and β forms of D glucose.

This reaction is comparable to the intramolecular formation of the 1.5 pyranose ring of the glucose, which results in the α and β forms.

If the aldehyde reacts with two molecules of alcohol of the same kind [reaction (c)], the resulting acetal is not asymmetric. If, on the other hand, the second alcohol molecule added to the hemiacetal is different reaction (c), the acetal is asymmetric. We shell see in the next section that such a reaction between two monosaccharides results in a glucosidic bond.

Glucose can be locked into its a and l configurations by methylation of the —OH at C-1. For example, when glucose reacts with methanol in acid solution two methyl glucosides are produced.

methyl-α-D-glucoside	methyl-β-D-glucoside
HC—OCH_3 (ring O)	H_3CO—C—H (ring O)
HCOH	HCOH
HOCH	HOCH
HCOH	HCOH
HC— (ring O)	HC— (ring O)
HCOH	HCOH
H	H

methyl-α-D-glucoside methyl-β-D-glucoside

Comparable reactions involve alcohols and ketones. Reaction (b) in Figure elsewhere in this chapter shows the formation of a hemiketal. It is the intramolecular version of this reaction which D-fructose undergoes to gain an additional asymmetric center at C-2 cause they are related to the heterocyclic parent compound furan. As shown in Figure, the designations α and β indicate the position of the —OH on C-2 relative to the ring oxygen. The common from of D-fructose is the P anomer (i.e., β-D-fructofuranose).

Conformation of the Monosaccharides

The pyranose and furanose forms of the sugars are frequently represented by *Haworth Projection* formulas, illustrated in Figure. The heavy lines in these structures represent the edge of the molecule nearest the reader.

Although the Haworth convention is convenient shorthand for indicating the general orientation of the rings and substituent groups, it does not allow an accurate representation of molecular conformation.

This is the most stable conformation for a cyclohexane type of ring structure. The C—OH bought C-2, C-3, and CL-4 are parallel to the nonadjacent side of the ring and are designated as equatorial ; the other bands are parallel to the axis of symmetry, more cut of the plane of the ring, and are called *Axial Bonds*. The equatorial -OH groups undergo

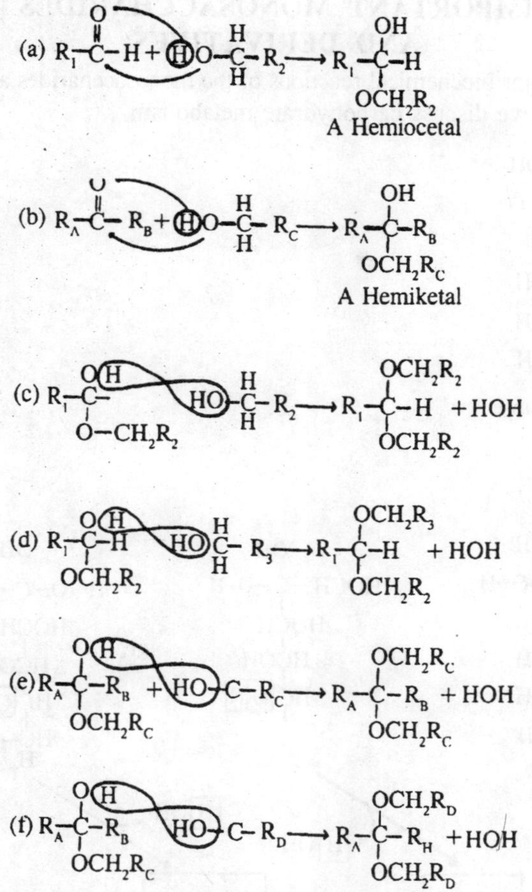

Figure 3.5 : Aldehyde hemiacetals and ketone hemiketals.

esterification more readily than the axial hydroxyls.

Actually, the pyranose ring of α-D-glucose exists in a "*chair form*" because of the normal valence angles of the carbons:

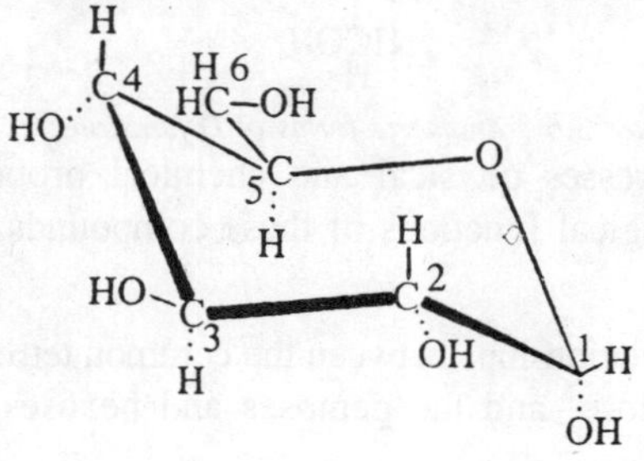

IMPORTANT MONOSACCHARIDES AND DERIVATIVES

The major biochemical reactions of the monosaccharides are considered where we discuss carbohydrate metabolism.

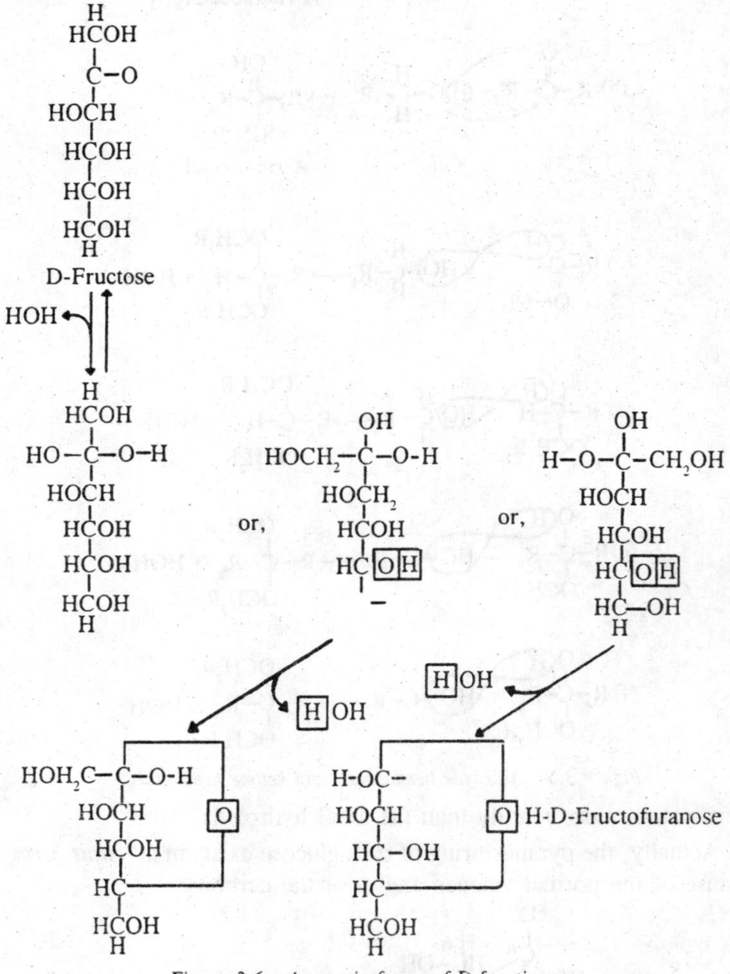

H
HCOH
C–O
HOCH
HCOH
HCOH
HCOH
H
D-Fructose
HOH
H
HCOH
HO–C–O–H
HOCH
HCOH
HCOH
HCOH
H
or,
OH
$HOCH_2$–C–O-H
$HOCH_2$
HCOH
HC OH
or,
OH
H–O–C–CH_2OH
HOCH
HCOH
HC OH
HC–OH
H
H OH
H OH
HOH_2C–C–O-H
HOCH
HCOH
HC
HCOH
H
O
H-OC–
HOCH
HC–OH
HC
HCOH
H
O H-D-Fructofuranose

Figure 3.6 : Anomeric forms of D-functions.

This section stresses physical and chemical properties that are relevant to the biological functions of these compounds.

Tetroses

The structural relationships between the common tetroses, erythrose, threose, and erythulose, and the pentoses and hexoses are shown in

Figure elsewhere in this chapter.

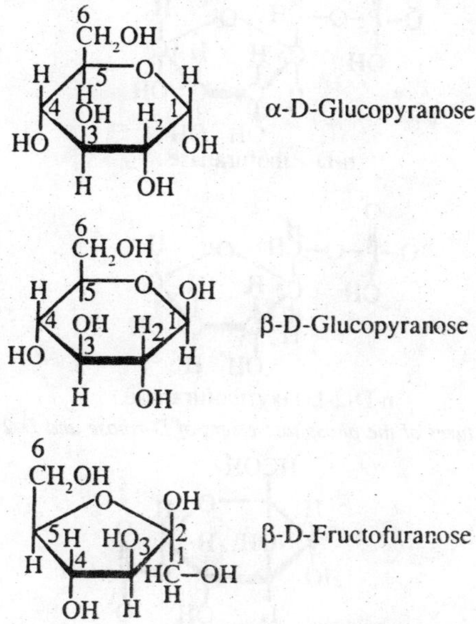

Figure 3.7 : Haworth projection formulas.

D-Erythrose 4-phosphate is a component of the transaldolase reaction of the phosphogluconate pathway.

Pentoses

Like D-erythrose-4-phosphate, the 5-phosphate derivatives of D-ribulose and D-xylulose are required in the phosphogluconate pathway. Two other important pentose derivatives are the 5-phosphate esters of D-ribose and D-2 deoxyribose, both of which are components of nucleic acids. The Haworth projection formulas are shown in Figure.

Hexoses

As in the case of the tetroses, and pentoses the phosphate esters of hexoses such as glucose, mannose, galactose, and fructose occurs as metabolic intermediates in a wide variety cells. Representative structures of the hexose derivatives areshown in Figure.

Hexose Alcohols

Reduction of the carbonyl group of the hexoses yields the sugar alcohols. Three important examples of this type of monosaccharide are D-mannitol, L-sorbitol, and myoinositol, whose structures.

Hexose Acids

An aldohexose such as glucose can have either or both of its

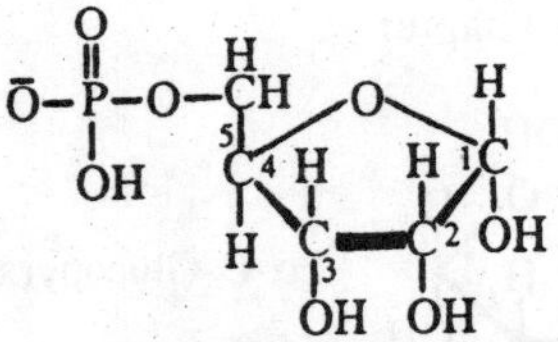

n-D-Ribofuranase

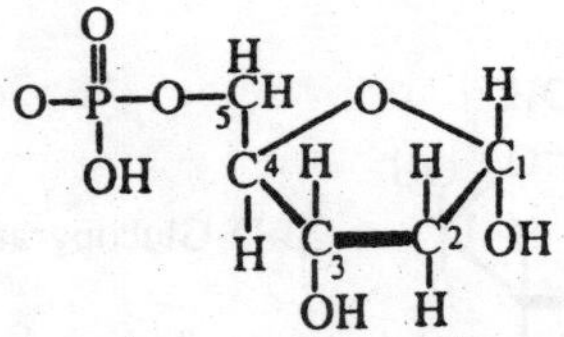

n-D-2-Deaxyribofuranose

Figure 3.8 : Structures of the phosphate esters of D-ribose and D-2-deoxyribose.

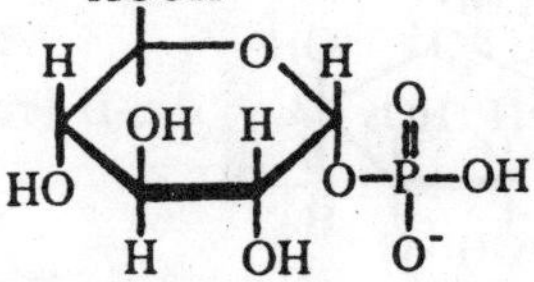

n-D-Glucose-1-phosphate

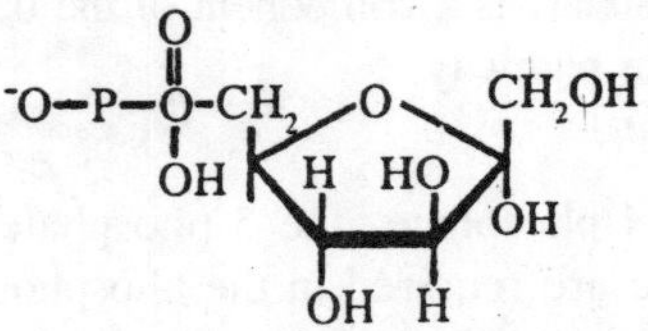

n-D-Fructose-6-phosphate

Figure 3.9: Examples of phosphate esters of hexoses.

terminal carbons oxidised to the carboxyl level. The structures and nomenclature of three types of hexose acids derived from D-glucose are presented. The phosphate ester of one of these acids, 6-phospho-D-gluconic acid, is an intermediate in the phosphogluconate pathway of glucose metabolism.

The "*onic*" acids, such as D-glucuronic acid, can constituents of *hyaluronic acid*, *chondroitin*, and other *mucopolysaccharides*. Both gluconic acid and glucuronic acid can form intramolecular esters, or lactones, which are five- or six-membered ring structures. Another aldonic acid important in hyuman metabolism is L-ascorbic acid, or vitamin C. L-ascorbic acid occurs as the γ-lactone and differs from other hexose acids in that it has an enediol linkage between C-2 and C-3.

Vitamin C an be reversibly dehydrogenated to dehydroascorbic acid. As will be described in more detail later, this compound is required for the establishment and maintenance of normal connective tissue structure and is also implicated in certain oxidation-reduction system.

Amino Sugars

Two hexose derivatives which are important constituents of polysaccharides and glycolipids in both *eukaryotic* and *prokaryotic* organisms are D-glucosamine and D-galactosamine. In both sugars the —OH on C-2 is replaced by an amino group as. These compounds occur principally in the N-acetylated form:

$$\begin{array}{c} | \\ H-N-\underset{\underset{O}{\|}}{C}-CH_3 \end{array}$$

Muramic and neuraminic acids are more complex forms of amino sugars which are essential structural components of cell walls of bacteria and of cell coats and membranes of eukaryotic cells. As shown with the formulas the carbon skeleton of each of these two sugar acids may be regarded as the union of a *hexose* and a *triose* (either pyruvic or D-lactic acid) ; the hexose backbone of N-acetylmuramic acid is D-glucosamine and that for N-acetylneuraminic acid is D-mannosamine. The more general term "*sialic acid*" is frequently used in identifying N-acyl derivatives of neuraminic acid.

PROPERTIES OF MONOSACCHARIDES

Physical Properties

Monosaccharides are sweet tasting colour-less solids, having solubility in water but sparingly soluble in alcohol and insoluble in ether. They have asymmetric carbon atom and therefore exist in different isomeric forms. An example is glyceraldehyde which may exist in two forms

$$\begin{array}{ccc} H & & H \\ | & & | \\ C=O & & C=O \\ | & & | \\ HO-C-H & & H-C-OH \\ | & & | \\ H-C-OH & & H-C-OH \\ | & & | \\ H & & H \end{array}$$

Asymmetric carbon atom

L-Glyceraldehyde D-Glyceraldehyde

which are mirror images of each other. If a polarised light (light vibrating in one plane) is allowed pass through a solution of these carbohydrates, the plane of e light gets rotated to either right or left. All *monosacchales* which have a D-*gly-ceraldehyde* unit are termed as sugars. L-monosaccharides have a L-glyceraldehyde unit their structure.

Carbohydrates isolated from living system are not pure ound, but a mixture of several isomers. The optical on may undergo a change due to interconversion of isomers. Fresh solution of glucose yields an optical rotation of +112° which gets changed to +52.7° on standing.

Similarly, fructose yields an optical rotation of -113° which gets changed to -92°. This change in optical rotation is termed as *mutarotation.*

Chemical Properties

As monosaccharides are aldehyde or ketone derivatives of the higher polyhydric alcohols, they give most, but not all, of the characteristic reactions of the carbonyl group as well as of the alcoholic group. In addition to these reactions, they also give rise to some special reactions due to the presence of a carbonyl and alcoholic group in the same molecule.

Reduction

The nature of the product formed by the reduction of aldoses depends upon the nature of the reagent. For example, D-glucose when reduced yields D-sorbitol and D-mannose produces D-mannitol. Sorbitol is found in the berries of many higher plants especially in the rosacea.

HOCH$_2$ / O / H H / OH H / HO / H OH — OH ⇌ CHO | H—C—OH | HO—C—H | H—C—OH | H—C—OH | CH$_2$OH $\xrightarrow{e^- \text{ i } H^+}$ CH$_2$OH | H—C—OH | HO—C—H | H—C—OH | H—C—OH | CH$_2$OH

When ketoses are reduced by the reducing agents, a mixture of two alcohols is obtained in each case. For example, D-fructose always yields a mixture of two epimers, D-sorbitol and D-mannitol.

Oxidation

As with reduction, the product of oxidation depends upon the nature of the reagent. Bromine water is a general reagent which selectively oxidises the —CHO group to a —COOH group and brings about conver-

$$\begin{array}{c} CH_2OH \\ | \\ C=O \\ | \\ HO-C-H \\ | \\ H-C-OH \\ | \\ H-C-OH \\ | \\ CH_2OH \\ \text{D-Fructose} \end{array} \xrightarrow{\sigma;\ H^+} \begin{array}{c} CH_2OH \\ | \\ H-C-OH \\ | \\ HO-C-H \\ | \\ H-C-OH \\ | \\ H-C-OH \\ | \\ CH_2OH \\ \text{D-Sorbitol} \end{array} \quad \begin{array}{c} CH_2OH \\ | \\ HO-C-H \\ | \\ HO-C-H \\ | \\ H-C-OH \\ | \\ H-C-OH \\ | \\ CH_2OH \\ \text{D-Mannitol} \end{array}$$

sion of an aldose to an aldonic acid.

$$\begin{array}{c} CHO \\ | \\ (CHOH)_n \\ | \\ CH_2OH \\ \text{aldose} \end{array} \xrightarrow{Br_2/H_2O} \begin{array}{c} COOH \\ | \\ (CHOH)_n \\ | \\ CH_2OH \\ \text{aldonic acid} \end{array}$$

Bromine water cannot oxidise the ketoses. When oxidation of aldoses is done with strong oxidising reagents like nitric acid, both the aldehyde and primary alcoholic groups will be oxidised to yield dicarboxylic acids known as aldaric acids.

$$\begin{array}{c} CHO \\ | \\ (CHOH)_n \\ | \\ CH_2OH \\ \text{aldose} \end{array} \xrightarrow{HNO_3} \begin{array}{c} COOH \\ | \\ (CHOH)_n \\ | \\ CH_2OH \\ \text{aldaric acid} \end{array}$$

Ketoses, on the other hand, are split up by the action of nitric acid to form a mixture of acids having fewer carbon atoms. However, it has been observed that the carbon chain of sorbose is not ruptured during oxidation with nitric acid, but a keto acid is the product.

$$\begin{array}{c} CH_2OH \\ | \\ CO \\ | \\ (CHOH)_3 \\ | \\ CH_2OH \\ \text{fructose} \end{array} \longrightarrow \begin{array}{c} CH_2OH \\ | \\ COOH \\ \text{glycollic acid;} \\ + \\ COOH \\ | \\ (CHOH)_2 \\ | \\ COOH \\ \text{tartaric acid} \end{array} \quad \begin{array}{c} CH_2OH \\ | \\ CO \\ | \\ (CHOH)_3 \\ | \\ CH_2OH \\ \text{sorbose} \end{array} \longrightarrow \begin{array}{c} COOH \\ | \\ CO \\ | \\ (CHOH)_3 \\ | \\ CH_2OH \\ \text{ketoacid} \end{array}$$

D-Glucose on periodate oxidation yields one mole of formaldehyde and 5 moles of formic acid.

HOCH$_2$ HOCH$_2$
H O OH H O H $H_2C=O$
H H
OH or OH + $5IO^-_4 \rightarrow$ + + $5IO^-_3$
O H HO OH 5HCOOH
H OH H OH
α-D-Glucopyranose α-D-Glucopyranose

Similarly, the glycoside methyl a-D-gluco-pyranoside will react with periodate as shown below:

6 HOCH$_2$ 6 HOCH$_2$
H 5 O H 5 O H
H 4 1 + $IO^-_4 \longrightarrow$ H 3
OH H CHO 1 + HCOOH + IO^-_3
OH 3 2 OCH$_3$ 4
H OH CHO OCH$_3$
2
Glycoside α-
D-Glucopyranoside

Benedict's reagent (an alkaline solution containing a cupric citrate complex ion) and Tollen's solution [$Ag(NH_3)_2OH$] oxidise and thus give positive tests with aldehydes and with ot-hydroxy ketoses. Both reagents, therefore, give positive tests with aldoses and ketoses. Sugars that give positive tests with Tollen's and Benedict's solutions are known as reducing sugars.

Ag^+ (complex) + aldose or ketose → Ag + Oxidation
Tollen's reagent *(Silve mirror)* *Products*

Cu^{2+} (complex) + Aldose or ketose → Cu_2O + Oxidation
Benedict's solution blue *(Brick red ppt)* *Products*

Oxidations with both Benedict's and Tollen's reagents take place in alkaline solution.

$$\begin{array}{l}CHO\\|\\(CHOH)_4\\|\\CH_2OH\end{array} + Ag_2O \xrightarrow[\text{tollen's reagent}]{} \begin{array}{l}COOH\\|\\(CHOH)_4\\|\\CH_2OH\end{array} + 2AG$$

gluconic acid silver mirror

$$\begin{array}{l}CHO\\|\\(CHOH)_4\\|\\CH_2OH\end{array} + 2Cu(OH)_2 \xrightarrow[\text{fehling's solution}]{} \begin{array}{l}COOH\\|\\(CHOH)_4\\|\\CH_2OH\end{array} + Cu_2O + H_2O$$

gluconic acid

Ester Formation

The formation of sugar esters indicates the presence of alcohol groups. For example, the total number of acy 1 groups which can thus be taken up by a molecule of the sugar is a measure of the number of such alcohol groups.

Glycoside Formation

This is one of the most important properties of monosaccharides. Consider as an example the formation of the methyl glycoside of glucose. When D-glucose in solution is treated with methanol and HCl, two compounds are obtained.

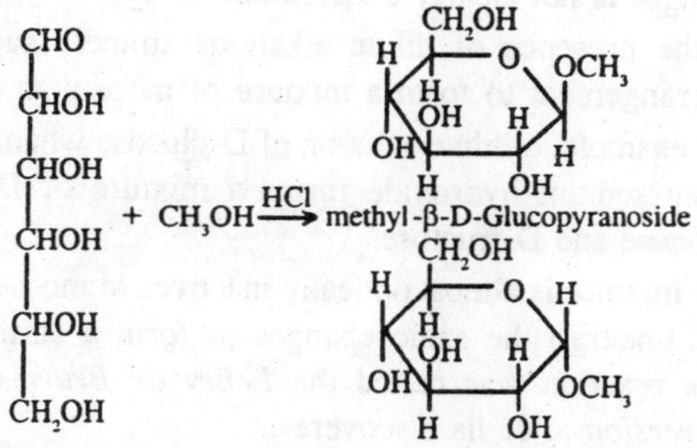

glucose + CH_3OH $\xrightarrow{HCl}$ methyl-β-D-Glucopyranoside

methyl-α-D-glucopyranoside

Glucosides are found in many drugs, spices and in the constituents of animal tissues. The glycosides which are important in medicines due to their action on the heart all contain steroids as the aglycone component. The examples of these include derivatives of digitalis and strophanthus.

Action of Acids

The monosaccharides are stable to the action of dilute (1.0 N) mineral acids but the concentrated acids decompose these compounds.

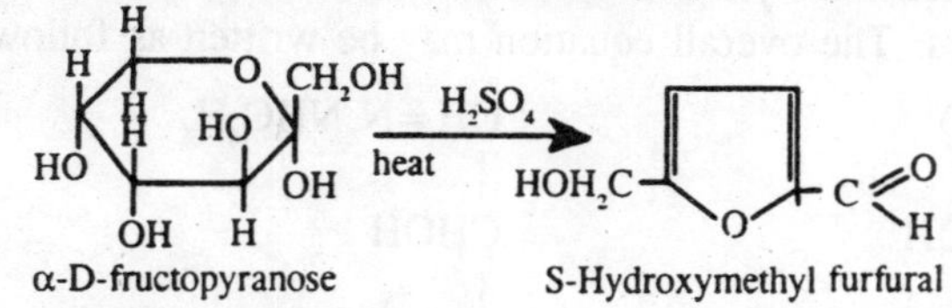

α-D-fructopyranose $\xrightarrow[heat]{H_2SO_4}$ S-Hydroxymethyl furfural

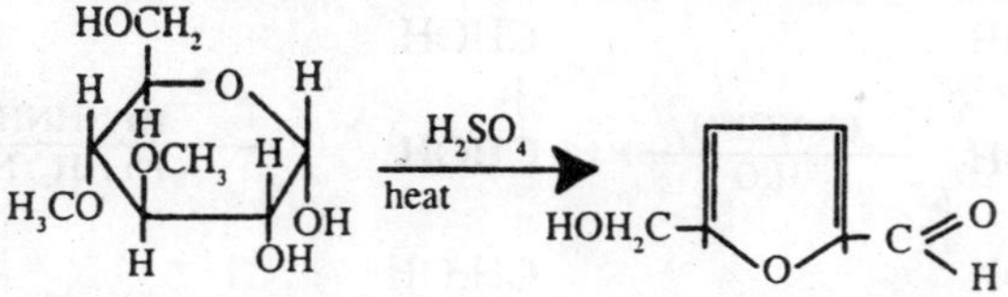

α-D-Glucopyranose $\xrightarrow[heat]{H_2SO_4}$ 5-hydroxymethyl furfural

However, aldohexoses and ketohexoses, when heated with strong mineral acids, are dehydrated and hydroxymethyl furfural is formed.

Action of Alkali

With alkali, monosaccharides react in various ways

(a) In concentrated alkali, sugar caramelises and produces a series of decomposition products.

Yellow and brown pigments develop, salts may form, many double bonds between carbon atoms are formed, and carbon-to-carbon bonds may rupture. However, the mechanism of these changes is not clearly understood.

(b) In the presence of dilute alkali or amines, sugars undergo rearrangement to form a mixture of more than one sugar.

For example, a dilute solution of D-glucose when warmed with dilute sodium hydroxide forms a mixture of D-glucose, D-mannose and D-fructose.

The mixture is almost optically inactive. Mannose and fructose also undergo the same changes to form a similar, mixture. This reaction was called the *Lobry de Bruyn-van Ekensein conversion* after its discoverer.

Osazone Formation

Aldoses as well as ketoses react with hydroxylamine and phenylhydrazine. With hydroxylamine the expected product is an oxime in aldoses and ketoses.

With phenylhydrazine, however, three moles of phenylhydrazine are consumed in aldoses as well as in ketoses to form *osazones*, *aniline* and *ammonia*.

Osazones are yellow crystalline solids and are used to characterise the sugars. The overall equation may be written as follows:

$$\begin{array}{c} CHO \\ | \\ CHOH \\ | \\ CHOH \\ | \\ CHOH \\ | \\ CHOH \end{array} \xrightarrow[H_2O]{Ph\ NHNH_2} \begin{array}{c} CH{=}N.NHC_6H_5 \\ | \\ CHOH \\ | \\ CHOH \\ | \\ CHOH \\ | \\ CH_2OH \end{array} \xrightarrow[-Ph\ NH_2,\ NH_3]{Ph\ NHNH_2}$$

Glucosone phenylhydrazone

$$\begin{array}{lcl}
CH{=}N.NHC_6H_5 & & CH{=}N.NHC_6H_5 \\
| & & | \\
C{=}O & & C{=}N.NHC_6H_5 \\
| & & | \\
CHOH & \xrightarrow[-H_2O]{Ph\ NHNH_2} & CHOH \\
| & & | \\
CHOH & & CHOH \\
| & & | \\
CHOH & & CHOH \\
| & & | \\
CH_2OH & & CH_2OH
\end{array}$$

Glucosone Phenylhydrazone — Glucosazone

$$\begin{array}{lclcl}
CH_2OH & & CH_2OH & & \\
| & & | & & \\
CO & & C{=}N.NHC_6H_5 & & \\
| & & | & & \\
CHOH & \xrightarrow[-H_2O]{Ph\ NHNH_2} & CHOH & \xrightarrow[-pH\ NH_2,\ -NH_3]{Ph\ NHNH_2} & \\
| & & | & & \\
CHOH & & CHOH & & \\
| & & | & & \\
CHOH & & CHOH & & \\
| & & | & & \\
CH_2OH & & CH_2OH & &
\end{array}$$

Fructose — Fructose Phenylhydrazone

$$\begin{array}{lcl}
CHO & & CH{=}N.NHC_6H_5 \\
| & & | \\
C{=}N.NHC_6H_5 & & C{=}N.NHC_6H_5 \\
| & & | \\
CHOH & \xrightarrow[-H_2O]{Ph\ NHNH_2} & CHOH \\
| & & | \\
CHOH & & CHOH \\
| & & | \\
CHOH & & CHOH \\
| & & | \\
CH_2OH & & CH_2OH
\end{array}$$

Fructosone phenylhydrazone — Fructosazone

When osazones are hydrolysed with hydrochloric acid, both phenylhydrazino groups are eliminated and osones are formed (osone is a dicarbonyl compound). For example, glucosazone, when hydrolysed with hydrochloric acid, yields glucosone.

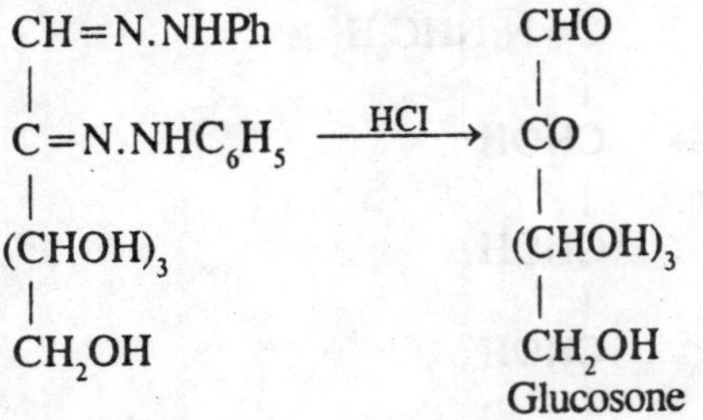

$$\begin{array}{l} CH{=}N.NHPh \\ | \\ C{=}N.NHC_6H_5 \\ | \\ (CHOH)_3 \\ | \\ CH_2OH \end{array} \xrightarrow{HCl} \begin{array}{l} CHO \\ | \\ CO \\ | \\ (CHOH)_3 \\ | \\ CH_2OH \end{array}$$

Glucosone

Addition of HCN

The general reaction for the addition of HCN to aldose may be represented as follows

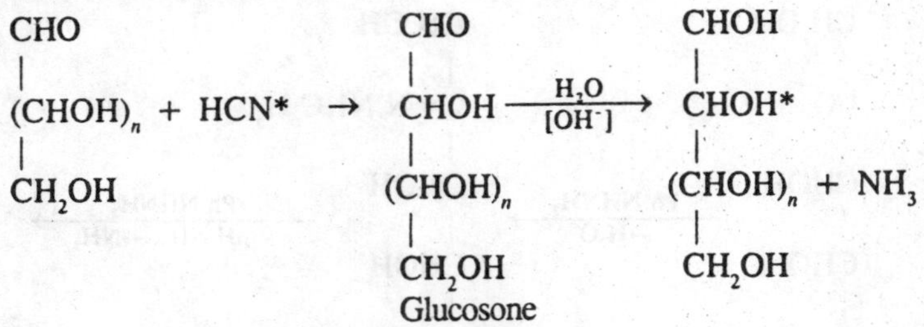

$$\begin{array}{l} CHO \\ | \\ (CHOH)_n \\ | \\ CH_2OH \end{array} + HCN^* \rightarrow \begin{array}{l} CHO \\ | \\ CHOH \\ | \\ (CHOH)_n \\ | \\ CH_2OH \end{array} \xrightarrow[{[OH^-]}]{H_2O} \begin{array}{l} CHOH \\ | \\ CHOH^* \\ | \\ (CHOH)_n \\ | \\ CH_2OH \end{array} + NH_3$$

Glucosone

*represents the asymmetric carbon atom.

The lactone of the sugars formed in the above reaction may in turn be reduced to the corresponding hemiacetal by *sodium amalgam*.

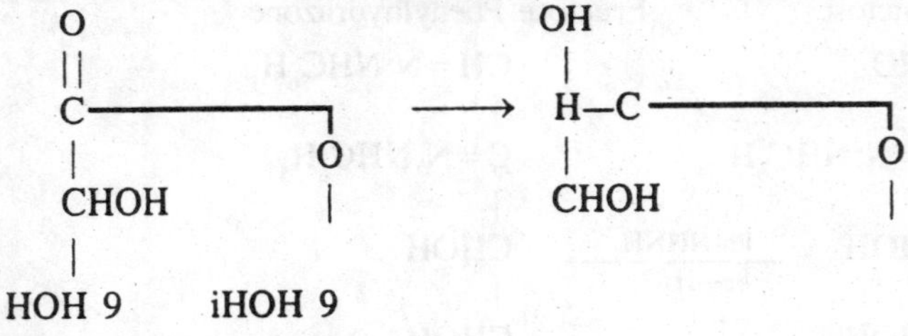

$$\begin{array}{l} O \\ || \\ C\text{———}\urcorner \\ |\quad\quad O \\ CHOH\quad | \\ | \end{array} \longrightarrow \begin{array}{l} OH \\ | \\ H{-}C\text{———}\urcorner \\ |\quad\quad O \\ CHOH\quad | \end{array}$$

HOH 9 iHOH 9

Phosphoric Acid Esters

In metabolism, phosphorylation of all sugars is the initial step in their metabolism. Thus, glucose is converted to glucose-6 phosphate. In most of the living systems, this compound is also converted into a-glucose 1-phosphate.

Glucose 6-phosphate is converted into fructose 6-phosphate which gets further phosphorylated to yield 1, 6diphosphate. The formation of these phosphoric acid esters occupies a unique formation in biochemistry.

The formulae of some phosphoric acid esters are given as follows:

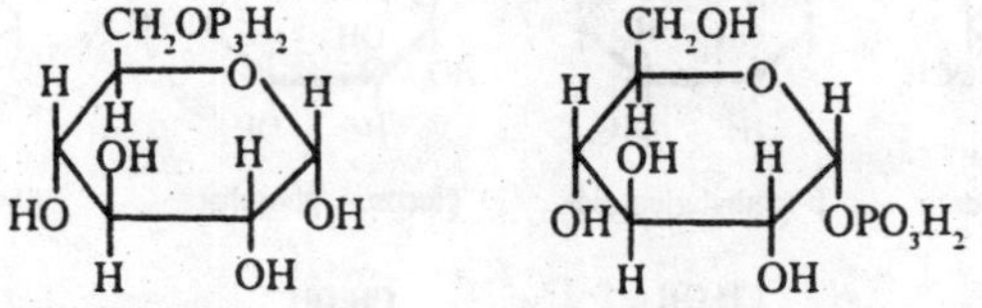

glucose 6-phosphate   glucose 1-phosphate

Reaction with Amines (N-Glucoside Formation)

When glucose is condensed with amines, *Schif's bases* are not obtain,a but instead N glycosides (analogous to the ordinary glucosides from alcohols) are obtained. For example, D-glucose reacts with *dimethylarine* in the presence ofan acid to form two isomeric N-glucosides.

Fermentation

When a solution of glucose is fermented by an enzyme called *zymase*, the products are mainly ethyl alcohol and carbon dioxide.

$$\underset{\text{D-Glucose}}{C_6H_{12}O_6} \xrightarrow{\text{zymase}} \underset{\text{Ethyl alcohol}}{2C_2H_5OH} + 2CO_2$$

Similar to glucose, fructose also undergoes fermentation by he *enzyme zymase* present in yeast to yield mainly *ethyl alcohol* and carbon dioxide.

$$\underset{\text{D-Fructose}}{C_6H_{12}O_6} \xrightarrow{\text{zymase}} 2C_2H_5OH + 2CO_2$$

Some Derived Monosaccharides

Derived monosaccharides differ from normal *monosaccharides* with respect to *aldoses* and *ketoses*. They include the *glycosides*, *sugar phosphates*, *gluconic acid*, *glucuronic amino sugars* and *vitamin C*.

Glucosides

The hydroxyl group on carbon 1 in the sugar molecule can be replaced by other redicals, forming compounds known as *glycosidei*. Thus if a *methyl group* replaces the *hydroxyl group* of glucose it results in the formation of *methyl glucoside*.

There are two methyl glucosides possible, α and β, corresponding to the α and β forms of glucose. Glucosides formed from *galactose* are known as *galactosides*. Among the complex glycosides occurring in nature are *digoxin*, which acts on the heart, and the antibiotic *streptomycin*.

Sugar Phosphates

C-1 or C-6 of glucose can react with *phosphoric acid* (H_3PO_4) to

α-methyl glucoside β-methyl glucoside glucose 1-phosphate glucose 6-phosphate

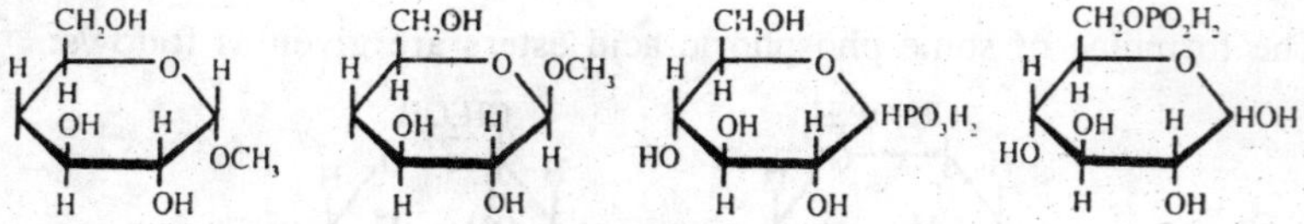

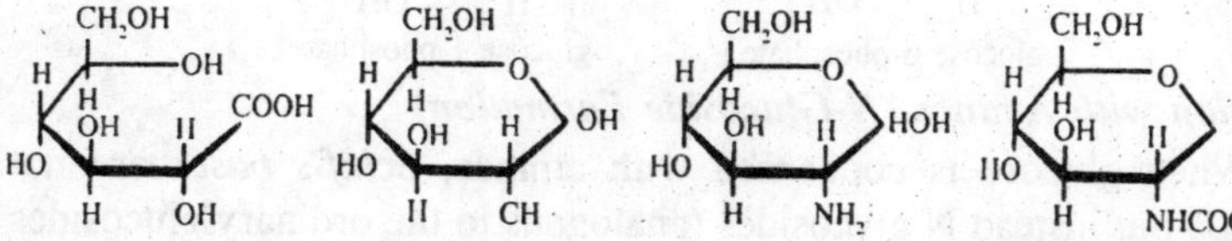

gluconic acid glucuronic acid glucosamine N-acetyl glucosamine

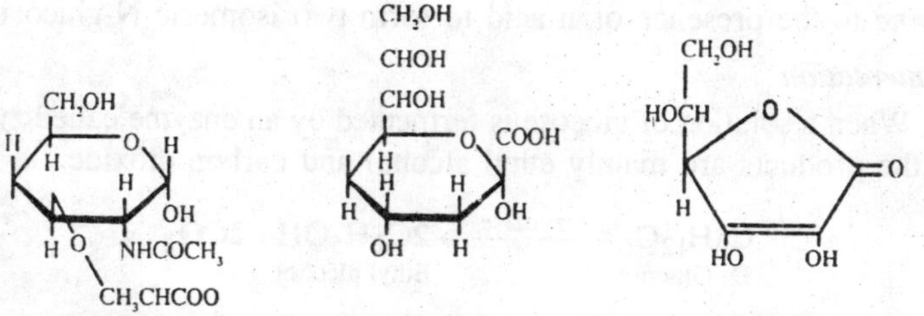

N-acetyl muramic acid neuraminic acid ascorbic acid

yield *glucose-1 -phosphate* and *glucose hosphate,* respectively. These compounds play an important *l* in carbohydrate metabolism.

Gluconic Acid and Glucuronic Acid

Oxidation of C-1 of glucose molecule to a *carboxyl group* yields *gluconic acid.* Production of a carboxyl group at C-6 yields *glucuronic acid.* e corresponding terms for any hexoses are *hexonic acid* and *hexuronic acid.*

Glucuronic acid is an important constituent of complex *polysaccharides* and is also an important coupling agent. Many drugs, pesticides, and environmental pollutants and hormones are coupled with glucuronic acid and excreted in urine or bile as *glucuronides.*

Amino Sugars or Hexosamines are formed when an amino sugar is introduced into hexoses. Thus glucose with NH_2 at C-2 forms *glucosamine.* This compound is extensively found in complex polysaccharides, usually in the form of its acetyl derivative *N-acetyl glucosamine. N-acetyl galactosamine*, the amino derivative of galactose, is found in *glycoproteins* and *glycolipids. N-acetyl muramic acid* is found in the polysaccharides of the cell wall of some bacteria. *Neuraminic acid* gives acetyl derivatives called *sialic acids.* These are constituents of some *glycoproteins* in bone and connective tissue and *glycolipids* of the nervous system.

Vitamin C (ascorbic acid) is a sugar acid of biological importance.

Oligosaccharides

Glycosidic Linkage

A sugar molecule can combine with an identical or a different type of a sugar molecule. The linkage between two monosaccharide sugar molecules is called a *glycosidle linkage* or *glycosidic band*. A sugar molecule has several reactive hydroxyl groups. The hydroxyl group of number one carbon atom (C-1 is known as the *glycosidic hydroxyl*.

It is very reactive and readily forms a *glycosidic link* with a hydroxyl group of another sugar molecule. When such a linkage occurs, one hydrogen atom and one hydroxyl group are eliminated to form H_2O. This process is one of candensation'and is in fact dehydration synthesis. The reverse process in which the molecule is cleaved with the incorporation of the elements of water is called hydrolysis. A linkage between C-1 of one monosaccharide unit and C-4 of another is called the 1, 4 linkage. Such a linkage occurs in disaccharides and in the unbranched chains of polysaccharides.

The 1, 4 linkage between two hydroxyl groups in the α position is called the α-1, 4 linkage (e.g., maltose). Similarly a linkage between two β hydroxyl groups is called the β-1, 4 linkage (e.g., in lactose). Linkage can also take place between C-1 of one monosaccharide unit and C-6 of another. Such a linkage occurs at the point of branching of a chain and is known as a 1, 6 linkage (e.g., in *isomaltose*).

The α or β configurations can be determined by specific enzymes. The enzyme maltase attacks only α-glycosides, while emulsin hydrolyses only P-glycosides. Starch has α-glycosidic bonds while cellulose has α-glycosidic bonds. The α bonds of starch can be cleaved by enzymes which do not cleave the β bond of cellulose. This explains why starch is a valuable food while cellulose has no food value for most animals.

Maltose and lactose both have unlinked potential aidehyde and are therefore reducing sugars. As long as the bond is not there between two C-1 atoms, a free aldehyde or ketone group is left on the disaccharide. The disaccharide therefore shows all the reactions associated with the groups. In sucrose linkage is through both the potentially reducing groups of the monosaccharides. Sucrose is therefore non-reducing.

Oligosaccharides are composed of 2 to 9 monosaccharide units or their derivatives. This commonly employed definition of oligosaccharides is an arbitrary one. In general it can be said that oligosaccharides contain fewer sugar residues than polysaccharides. The oligosaccharides are named according to the number of monosaccharide units they contain. Thus *disaccharides* contain two monosaccharide units, *trisaccharides* three, *telrasaccharides* four, *pentasaccharides* five, etc.

DISACCHARIDES

These are formed by union of two monosaccharides. They are united by linkage between the first carbon of one monosaccharide with the second or the fourth carbon of another monosaccharide by what is known as the glycosidic linkage. The structures of the three biologically important sugars—*maltose*, *lactose* and *sucrose* are shown in Figure elsewhere in this chapter.

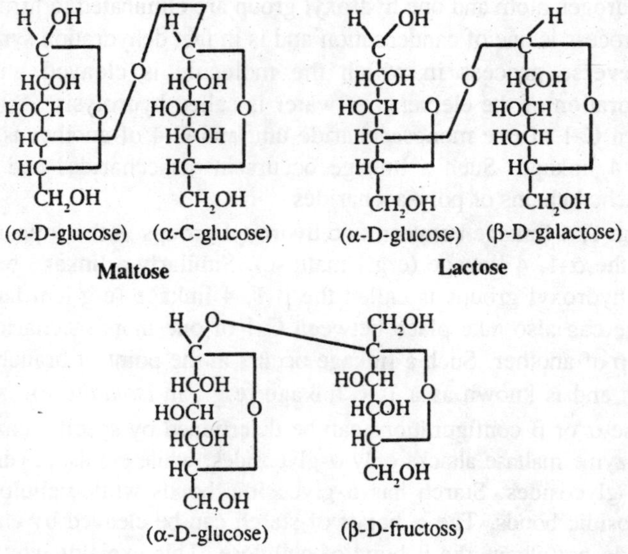

(α-D-glucose) (α-C-glucose) (α-D-glucose) (β-D-galactose)

Maltose **Lactose**

(α-D-glucose) (β-D-fructoss)

Sucrose

Figure 3.10 : Disaccharides.

They are colourless crytalline substances readily soluble in water and sweetish to taste.

Maltose

This is made up of two alfa D-glucose units united by the 1, 4-glycosidic linkage. Since one of the aldehyde groups remains free, it exhibits all the properties of the aldehyde group such as formation of osazones, reduction of cupric salts and others. Maltosazone crystals have a charac teristic petal like appearance and a cluster of them looks lik the sunflower. Maltose is formed during the digestion of starches by enzymes or by dilute acids.

Lactose

This is made up of a molecule of alfa D-glucose united by 1, 4-glycosidic linkage to a molecule of beta D-galactose. The aldehyde group of glucose remains free and exhibits the properties due to that group. I actosazone crystals have a typical hedgehog shape and are readily identifiable. Lactose is present in milk (*milk sugar*) and by hydrolysed to glucose and galactose by the enzyme '*lactase*' present in intestinal juice,

Sucrose

It is made up of one molecule of alfa-D-glucose and one of beta D-fructose united by a glycosidic linkage between the aldehyde and keto groups (C_1 of glucose and C_2 of fructose). There is hence no free aldehyde or keto group in sucrose. It is a non-reducing sugar. It does not also form *osazones*. Solutions of sucrose exhibit a specific rotation of + 62.5°. On hydrolysis with acid or with the specific intestinal enzyme '*sucrase*', it is converted into an equimolar mixture of glucose and fructose whose specific activities are +52.5° and -92°.

The mixture exhibits a net specific rotation of -19°. The phenomenon by which the, dextrorotatory sucrose is converted to a *levorotatory* mixture of glucose and fructose is known as '*inversion*'. Sucrose is also called the '*invert sugar*' and the enzyme sucrase is also known as '*invertase*'. Sucrose occurs widely in nature as cane sugar, beet sugar and in several ripe fruits.

Cellobiose (from cellulose) and gentiobiose are other disaccharides of glucose formed by β 1-4 and β 1-6 glycosidic linkages. Both are reducing sugars. Trehalose is a disaccharide formed by two glucose units combining by 1-1 linkage. It is a non-reducing sugar. It is contained in the hemolymph of some insects.

Trisaccharides

Raffinose (galactose +glucose+ fructose) is found in sugar beets. Melezitose (glucose+ fructose+ glucose) is found in the sap of some coniferous trees.

Polysaccharides

The term polysaccharide is usually employed for polymers continuing at least 10 monosaccharide units. Polysaccharides are named after their monosaccharide monomers by changing the—*ose* ending of the monosaccharide to-*an*. Thus *cellulose*. a polysaccharide formed of *glucose* units, is called *glucan*. *Fructose* polymers are called *fruetans* (previously *fructosans*) and *perrtose* polymers *pentans* (formaly *fructosans*). Since

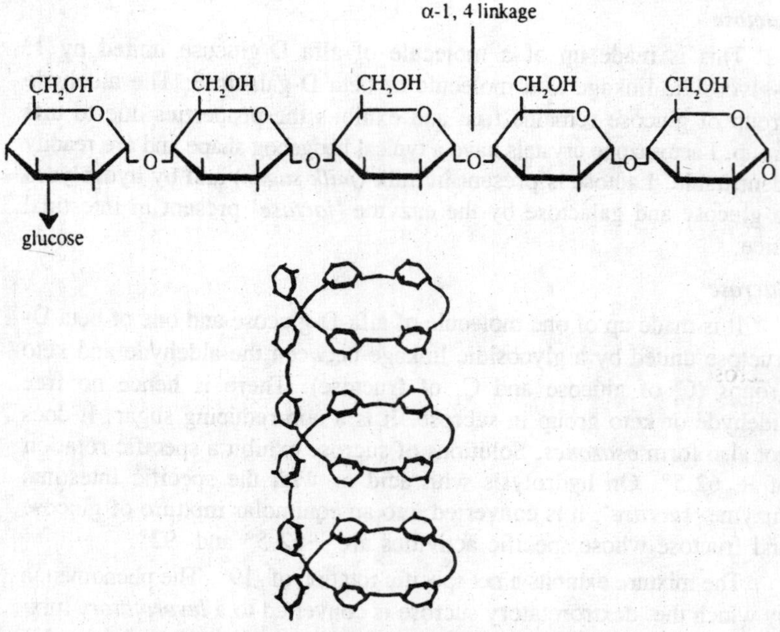

Figure 3.11 : (A) Structure of amylose showing α-1,4 linkage.
(B) Helical coil of amylose on suspension in water.

the term *glycose* is used for any monosaccharide, the general name for any polysaccharides is *glycan*. If a polysaccharide contains more than one type of sugar units the two sugars are named in alphabetical order. Thus, a polysaccharide consisting of *D-glucose* and *D-mannose* units is termed *D-gluco-D-niannoglycan*.

Sine old maines which have not been changed to, conform with the-an ending include ***starch***, ***cellulose***, ***pectin***, ***amylopectin***, ***inulin***, *chitin* and *liquirin*. Polysaccharides have been classified as *homopolysaccharides (monncglycans)* and *heteropolysaccharides* (*heterogly-cans*) *Homoglycans* contain only one type of monosaccharide while *heteroglycans* contain atleast two types. Most polysaccharides contain one or two types of glycose units.

Polysaccharides containing three types of glycose units are less common. The term *homoglycan* is used when at least 9% of the sugars are of one type, because it is very rarely that a polysaccharide can be obtained in pure form. Polysaccharides are formed by condensation of many molecules of monosaccharides with corresponding elimination of water molecule. The monosaccharide units are joined by glycosidic bonds.

A. Homopolysaccharide (homoglycans)

Some of the better known homoglycans are:

Glucans (of *glucose* monomers)—strach, glycogen, cellulose, chitin.

Galacrans (of *galactose* monomers)—agar, pectin galactan, from snails.

Mannans (of *mannose* monomers) —yeast mannan.

Xylans (of *xylose* monomers)—hemicellulose xylan.

Fructans (of *fructose* monomers)—inulin.

Strach

Strach is the reserve substance in plant cells. It is a polymer of *D-glucopyronose* units linked by α-1-4. It -consists of a mixture of *amylose* and *amylopectin* in the proportion of 1 to 4. Both are high molecular weight compounds. *Amylose* is *linear* while *amylopectin* is branched. Treating strach with hot water dissolves amylose, while amylopectin remains.

Amylose contains about 200-500 glucose units which are arranged in the form of a straight chain. Its molecular weight varies from a few thousand to about 150,000. Amylose gives intense blue colour with iodine.

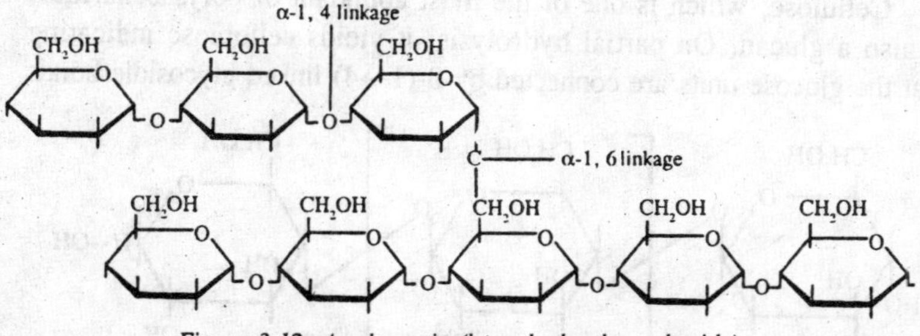

Figure 3.12: Amylopectin (branched polysaccharide).

Amylopectin has over 1,000 glucose residues and its molecular weight is about 200,000 to 1,000,000. It has a branched structure. Amylopectin consists of glucose units linked by α-1,4 glycosidic bonds. There are occassional α-1,6 glycosidic bonds which cause branching.

Strach on hydrolysis yields lower molecular weight polysacharides, and finally glucose or maltose. Partial hydrolysis results in substances called *dextrins*. At first *erythrodextrins* are formed which give red colour with iodine. Further hydrolysis results in the formation of *achroodextrins* which do not give the red colour with iodine. Finally reducing sugars appear.

Dextrins are easily digested and are therefore used for feeding infants. When mixed with milk they prevent formation of curds in the stomach. The enzymes which bring bout hydrolysis of starch are called *amylases*.

Glycogen is the major reserve carbohydrate in animals and is therefore also called *animal starch*. It is found mainly in the *liver* and in *muscles*. The glycogen in the *liver* supplies glucose to all tissues through the blood. *Muscle glycogen* on the other hand is available during contraction of muscle.

Chemically glycogen is analogous to strach. Both are glucans consisting of glucose units linked by α-1,4 glycosidic bonds, with branch chains formed at α-1,6 bonds. They differ mainly in their molecular weights and degree of branching. Molecular weights range from about 300,000 to 100 million.

The branching chain of glycogen average about 12 glucose units as compared with the 25 or so for starch. Glycogen has branch points about every 8-10 glucose units. About 5,000 to 15,000 glucose units make up glycogen. Glycogen is nonreducing and gives a red colour with iodine.

CELLULOSE AND OTHER CARBOHYDRATES

Cellulose, which is one of the most abundant of polysaccharides, is also a glucan. On partial hydrolysis, it yields cellobiose indicating that the glucose units are connected by β-(1→4) linked glycosidic bonds

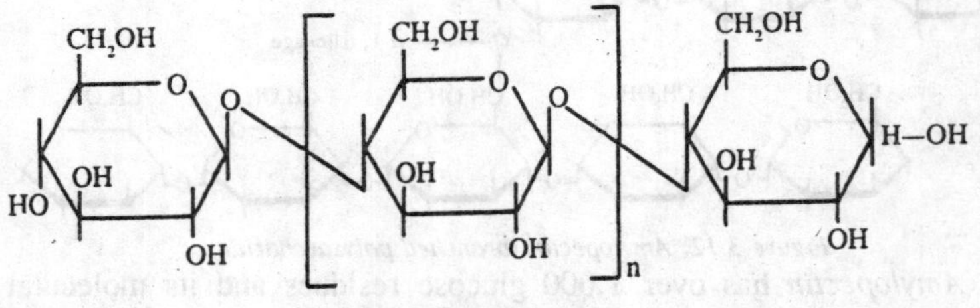

CH_2OH CH_2OH CH_2OH
O O O O O
OH OH OH H—OH
HO
OH OH OH
n

Figure 3.13: Cellulose structure.

in as straight polysaccharide chain. Cellulose, like cellobiose, is nutritionally useless to man as there is no enzyme in the human body which can act upon a β-glucosidic linkage. Ruminants, however, digest cellucose with the aid of symbiotic microorganisms which can attack β-glucosidic links and break down the glucose to organic acids.

Plant Cell Wall Polysaccharides

The plant cell is distinguished from the animal cell in having a rigid cell wall composed mainly of cellulose and by the presence of a large fluid-filled vacuole in the mature cell. Most cell walls are joined together but during the processes of growth and differentiation, cells elongate as well as multiply. Meristematic cells associated with these processes have initially a thin plasma membrane which is freely permeable to water, ions and small molecules.

The rate of entry of these materials determines the osmotic pressure within the cell which in turn controls the growth of the cell and the formation of the cell wall. The thick secondary cell wall which is formed later by secretion confines the enlargement and multiplication to one axis. The rigidity and strength of the cell wall are due to the extensive cross-linking of cellulose fibres with other complex polysaccharides like *xyloglucan*, *arabinogalactan* and *rhamnoglucuronen*.

Each cellulose fibre appears to be surrounded by a layer of xyloglucan. The walls contain other carbohydrate material like hemicelluloses and pectins and non-cellulosic substances like lignin and extensin. *Lignins* are polymeric non-carbohydrates containing parahydroxy phenyl propanes as the repeating units and representing about 15-30 per cent of the dry weight.

They are probably derived from *shikimic acid,* which may be a precursor of the phenolic residues of lignins. *Extensin* is a complex glycoprotein attached covalently to the cellulose fibrils. It is composed of glucose and galactose residues and resembles collagen in its high content of hydroxyproline.

OH
HOOC— —OH
OH

Mannan composed of mannose units is present to the extent of 16 per cent in baker's yeast.

Inulin occurs in dahlia bulbs and Jerusalem artichoke and is predominantly a fructosan which yields D-fructose on hydrolysis and a small number of D-glucose units. It is a linear polymer consisting of about 33 fructofuranose units joined together by means of β-(2→1) glycosidic linkages and having a molecular weight of around 5,200. It is an elongated molecule which is able to pass the glomerular filter of the kidney and is not secreted by the kidney tubules. It has, therefore, been used to measure the rate of glomerular filtration by injection into the blood stream as it is not metabolised and is eliminated at a constant rate by the kidney.

Agar

Natural agar from seaweeds consists mainly of the calcium salt of a sulphuric ester of a galactan. It is used as a gelforming agent in media for culturing micro-organisms. It is not liquefied by organisms that digest gelatin. Agar is nondigestible and is sometimes used to provide bulk to the faces in the treatment of constipation.

Most seaweed polysacc-harides are homoglycans. Alginic acid which is a linear polymer of D-mannuronic acids linked by β-(1→4) bonds has found industrial and food uses. It is used as a stabiliser of ice cream, an emulsifier in foods and salad dressings and even as an edible sausage casing.

Pectic substances are widely used in the food industry to form gels with sugar and fruit acids. They correspond to the hyaluronic acid of the ground substance of animal tissues. They are composed of three polysaccharides, a galactan, the methylester of a galacturonan and an araban. Mucilages and gums are even more complicated plant polysaccharides containing uronic acids as well as hexoses and pentoses.

Hernicelluloses

These are related substances that contain more than one sugar (glucose, galactose, xylose and arabinose) and are extractable from the cell walls of plants by alkaline solutions (17.5 per cent sodium hydroxide). Cellulose swells but does not dissolve in this solution. The celluloses along with the *hernicelluloses* form a good part of the roughage or "*bulk*" in the human diet.

Chitin is the structural material of the exoskeletons of insects and the shells of crustaceans and is a polymer of N-acetyl-D-glucosamine connected by β-(1→4) glycosidic bonds.

Mucopolysaccharides (Glycosaminoglycans)

The prefix "*muco*" was used to indicate their relationship to mucus and viscid secretions in general and this terns has been loosely used to cover a large group of complex carbohydrates and carbohydrates attached to proteins or lipids (*mucoproteins*, *mucolipids*), all of which are characterised by high viscosity in solution.

For convenience, however, the term mucopolysaccharide may be used to denote a complex of carbohydrate and protein whose reaction is predominantly polysacharidic and is applied to heteroglycans which contain generally residues of both a uronic acid and a hexosamine (glucosamine, 2-amino-2-deoxy-D-glucose; and galactosamine (chondrosamine) 2-amino-2-deoxy-D-galactose). Jeanloz has suggested the name *glycosaminoglycans* to describe this group of substances. They are further subdivided into a acidic mucopolysaccharides which contain hexosamine and, usually, uronic acid with or without sulphate and neutral mucopolysaccharides which contain hexosam Inc and a neutral monosaccharide. The acid mucoplysaccharides include the sulphatefree hyaluronic acid and the sulphate-containing isomeric chondroitin-4- and 6-sulphates, heparin, dermatan sulphate (chondroitin sulphate B, β-heparin), heparin sulphate (heparin sulphate, hepearan sulphate), hyaluronic acid sulphate, and keratosulphate (without uronic acid).

In most connective tissues of animals, the acidic mucopolysaccharides are complexed with protein or peptide residues. They are all negatively charged and all the well-known members of the group are linear polymers which behave as anionic polymeric electrolytes. The isolation of pure entities and the avoidance of artefacts depends on the delicacy of the methods employed. One of the important components of mucoproteins, neuraminic acid, is attached by a glycosidic link that is far more labile than those linking the other monosaccharide units.

This linkage can be preferentially hydrolysed by acid or the specific enzyme neuraminidase with little alteration or damage to the remaining structure. The use of detergents like cetylpyridinium chloride and cetyl trimetbylammonium bromide has enabled the infact polysaccharide-protein complex to be isolated in a number of cases, the large positively charged ions complexing with the negatively charged mucopolysaccharide. Such complexes, though insoluble in water, show a gradation of solubilities in aqueous solutions of varying ionic strength and are sometimes soluble in organic solvents.

Selective precipitation of acid polysaccharides can also be achieved when acid conditions are used to suppress the ionisation of carboxyl and sulphate groups. Methods for the specific splitting of mucopolysaccharides into the component carbohydrate and protein moieties are being developed with use of specific enzymes a hydrolytic agents.

Biological Function

The mucopolysaccharides are extra-cellular in location and are associated with connective tissue in different parts of the body. Three major cytological components are generally recognised in connective tissue: Cells, extracellular fibres and extracellular amorphous ground substance. Besides the fundamental similarly in the function of connective tissue to bind and support the various organs of the body, there appears to be present also a chemical similarity in the connective tissues of various parts of the body in their content of *mucopolysaccharides* and *mucoproteins*.

The acid mucopolysaccharides are found in the amorphous ground substance which forms the cement that holds the cells together and comprises the medium in which the cells move and the moving parts are lubricated. By reason of its content of surfaceactive components, variations in the state and composition of the ground substance modify considerably the activity and permeability of cell membranes and, therefore, of cells and tissues.

It has been suggested that tissue mucopolysaccharides might effects and modify various physiological and pathological processes like movement of water and electrolytes, joint lubrication, blood coagulation, fertilisation, hair growth, wound healing, calcification and so on. Among the several proteins associated with mucopolysaccharides may be mentioned the mucoproteins orosomucoid (α_1-globulin), haptoglobin (α_2-globulins), and the gamma globulins found in plasma, the glycoprotein hormones of the anterior pituitary gland which include the gonadotropins, thyrotropin, and prolactin, mucin from the submaxillary gland and proteins associated with the blood group substance as well as ovomucoid which is known to be an inhibitor of pancreatic trypsin.

The blood group substances are widely distributed in the body and are present in saliva, gastric juice, other tissue extracts and also in ovarian cyst fluid. More than sixty different antigens have been identified with specific blood groups in man but haemolytic disturbances are associated primarily with the *D-antigen* of the *Rh group* and with incompatibility of ABO factors. *Blood group antigens* are glycoproteins or glycolipids. The ABO factors are associated with the transmembrane *Band 3 protein* of red blood cells.

The synthesis of these antigens is determined by specific genes termed H, A, B, and so on. Gene H is concerned with the attachment of fucose to the I carbohydrate moiety consisting of galactose and N-acetyl glucosamine residues already present on the protein to establish blood group O. The transferase dependent on gene A permits attachment of N-acetyl galactosamine to form group A antigen.

The transferase dependent on gene B attaches galactose which determines group B antigen. The synthesis of the antigens is completed by adding several units of fucose, the position of these fucose units determining the specificity of the blood group. The protein collagen is embedded in a mucopolysaccharide matrix.

The ability of several mucoproteins to inhibit the clumping of red cells, induced by heat-treated influenza virus, appears to be associated with the sialic acid portion.

Chemistry of Mucopolysaccharides

The most abundant of the group of acid mucopolysaccharides is probably *hyaluronic acid*, the ground substance of mesenchyme. It has been isolated from the skin, synovial fluid, vitreous humour, Wharton's jelly obtained from the umbilical cord, and also from haemolytic streptococci. A hyaluronic acid-protein complex has also been isolated from human synovial fluid using zone electrophoresis. It is a high molecular weight material usually polydisperse, ranging in molecular weight from 50,000 to many millions depending on the method of isolation.

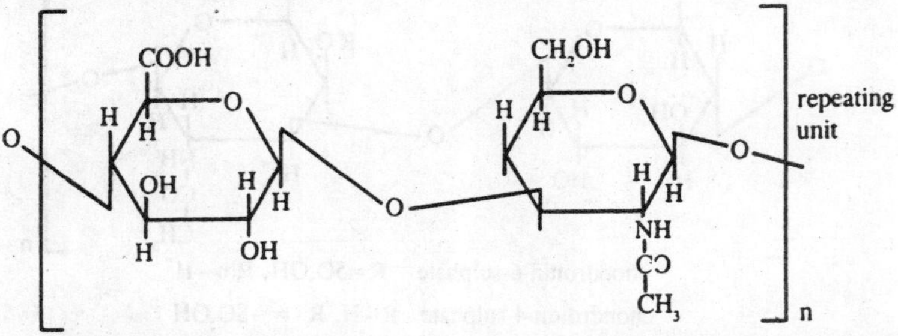

COOH CH$_2$OH
O H H H H repeating unit O
OH H H O H H NH
H OH CO
CH_3 n

The large size of the molecule and the nature of the constituent units bestow a high viscosity on solutions of the polymer constituted of the disaccharide, hyalobiuronic acid. This disaccharide is composed of N-acetyl-D-glucosamine and D-glucuronic acid in combination.

Hyaluronic acid appears to bind water in the tissue spaces and in the skin. For a pure liquid and for many solutions, the viscocity is generally independent of the velocity gradient and this is referred to as "*Newtonian viscosity*". Asymmetric molecules, however, tend to take up an average position with their longer dimensions more or less oriented with the direction of flow. In this position they make less contribution to the viscosity than when they are randomly oriented.

The viscosity of such solutions tends to fall as the velocity gradient increases and this is termed *"non-Newtonian viscosity"*. Such viscosity changes make synovial fluid an ideal lubricant in joints which have to bear weight as well as facilitate movement. These viscosity changes have been attributed to the presence of byaluronic acid. *Hyaluronidase* is the enzyme that catalyses the depolymerisation of hyaluronic acid and by reducing its viscosity facilitates the diffusion of material into tissues.

The enzyme is persent in spermatic fluid and is usually prepared from bull or ram tests. It is found also in pneumococci and streptococci, snake and bee venoms, skin autolysates and extracts of leech heads. The testicular preparation cleaves the hyaluronic acid to the *tetrasaccharide* stage and' affects only the glucosaminidic linkages between C_1 of the glucosamine moiety and C_4 of the glucuronic acid, the uronidic linkages not being attacked. The leech enzyme hydrolyses the glucuronidic bond. The bacterial hyaluronidase cleave the tetrasaccharide obtained by mammalian hyaluronidase activity into two disaccharides.

Many micro-organisms produce the enzyme and the presence of the enzyme may facilitate spread of infection and movement of bacilli into

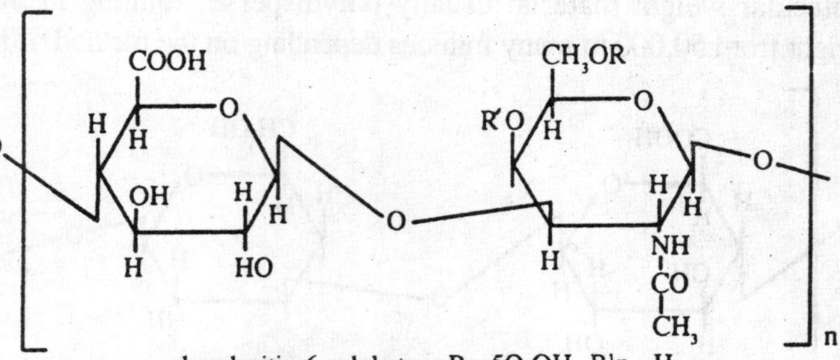

chondroitin-6-sulphate $R=SO_2OH$, $R'n—H$

chondroitin-4-sulphate $R—H$; $R' = -SO_2OH$

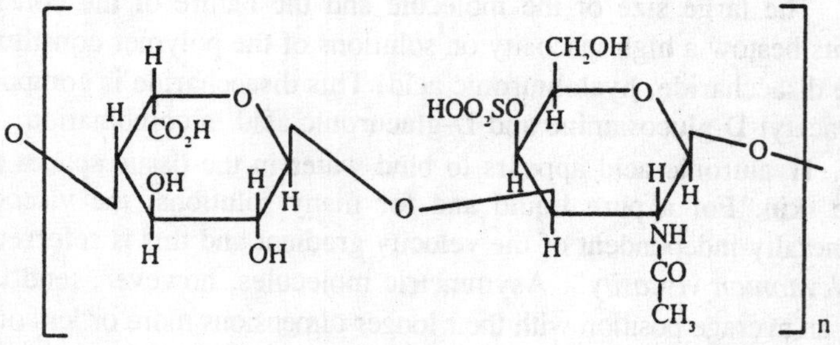

repeating unit (1→4) -O-α-L-idopyronaosyl uronic acid (1→ 3)-2-acet-amido-2-deoxy-4-O-sulpho-β-D-galoctopytanose

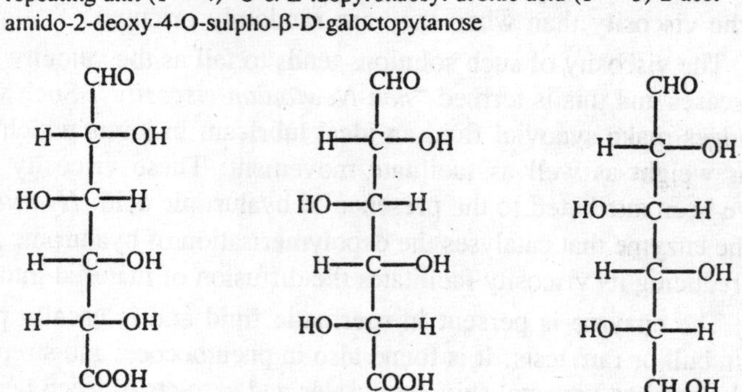

D-glucuronic acid

L-iduronic acid

L-idose

the surrounding tissues. Purified enzyme preparations are used in clinical medicine to facilitate diffusion of fluids and drugs when administering fluids by the subcutaneous route in states of dehydration or in producing local anaesthesia. It is claimed that it helps the removal of adhesions encounted during surgery.

Chondroitin-4- and 6-sulphates (Chondroitin sulphates A and C) occur in cartilage and adult bone and as minor constituents in other tissues.

The linkage is assumed to be the same in both mucopolysaccbarides in view of the fact that the rate of breakdown by bacterial or animal beta-hexosaminidases is simila. These componnds are associated with collagen and may account for the physical characteristics of the cartilaginous tissue in which they are present.

Dermatan sulphate β-heparin, chondroitin sulphate B). It has been isolated from tendons, aorta, heart valves, sclera, lung parenchyma and is present in large amounts in pig skin. It is distinguished from the cbondroitin-4- and 6-sulphates in not being degraded by testicular hyaluronidase. Acid hydrolysis of the compound yields D-galactosamine, acetic acid, sulphuric acid and a uronic acid which is believed to be L-iduronic acid which is an epimer of D-glucuronic acid at the fifth carbon from the aldehydic group.

Keratan sulphate (Keratosulphate)

It occurs fairly widely in connective tissues and has been isolated from bovine cornea. It is roported in the nucleus pulposus, the wall of the aorta and costal cartilage. It is composed of N-acetyl-D-glucosamine, D-galactose and sulphate and contains no uronic acid residue. A small amount of L-fucose has also been reported. There appears to be a similarity of structure between keratosulphate and the blood group substances which yield the same products on hydrolysis except for the polypeptide protein.

Heparin

Originally obtained from the liver, it is present also in the lung, the walls of large arteries and in other areas where mast cells are found. It is an anti-coagulant *in vitro* and *in vivo*. On hydrolysis it yields glucuronic acid, glucosamine, acetic acid and sulphuric acid.

The presence of iduronic acid in the molecule has been reported. The sulphuric acid is present as ester sulphate in glucuronic acid as well as in amide linkage with the amino group of the glucosamine to form sulphamic acid-$NHSO_2OH$ groups with an additional O-sulphate at

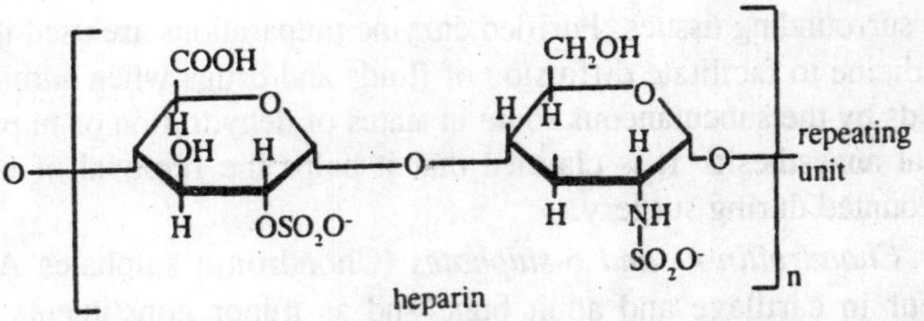

heparin

C3. The D-glucuronic acid (L-iduronic acid) has an O-sulphate linkage probably at C2.

Heparitin sulphate has been isolated from normal human and cattle aorta, from urine, liver and spleen of patients with gargoylism and also from amyloid liver. Heparitin sulphate differs from heparin in that a variable fraction of the amino groups bears an N-acetyl residue instead of a sulphate.

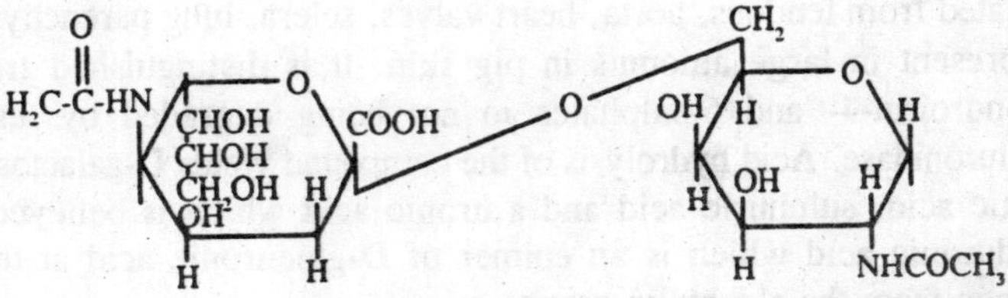

Salivary mucin (mucus) contains numerous molecules of the disaccharide unit 6-(N-acetyl-a-neuraminyl) N-acetyl-Dgalactosamine as the prosthetic group in association with protein. Other components of mucin (mucus) are L-fucose, D-galactose and N-acetyl glucosamine. This material is present throughout the gastrointestinal tract.

It has a high water content (around 95 per cent) and is associated with a large glycoprotein (about 2 million daltons). The carbohydrate chains are arranged on the glycoprotein like bristles on a testtube brush and are able to enmesh large quantities of water and foreign matter. Prostaglandins stimulate mucus secretion.

Bacterial Cell Wall Polysaccharides

The chemical composition of the bacterial cell wall has been worked out in great detail since it is concerned with the virulence of bacteria, the presence of specific antigens on the cell surface and,, the mode of action of some antibiotics, like penicillin, which prevent the synthesis of the cell wall.

The Danish physician Gram used gentian violet followed by iodine hoping to stain the nuclei of kidney cells violet and the tubules brown. After this treatment, the section was rapidly decolourised by alcohol.

Certain bacteria present in the section, however, remained blue-black and this chance observation led to the development of a stain which could distinguished bacteria into Gram-positive bacteria and cocci (like the tetanus bacillus, staphylococci and streptococci) and Gram-negative bacteria and cocci like *E. coli* and the *gonococcus*.

Ehrlich's great contributions to chemotherapy were based on the postulate that bacteria could be attacked by a form of differential staining or toxicity which mould not affect the host. Bacteria are surrounded by a tough envelope which preserves their integrity and shape. When Gram-positive bacteria are subjected to the action of lysozyme, an enzyme present in nasal secretions, tears, saliva and egg white, they swell and rupture of the cell membranes takes place. Such rupture is prevented by the presence of isotonic sucrose in the medium, though breakdown of the cell wall occurs, and the bacterial cell covered by a delicate thin membrane (protoplast) is left behind.

Gram-positive bacteria and cocci are composed of complex polymers with interwoven strands of carbohydrate and amino acid residues, some of them being D-amino acids. The carbohydrates include as major components N-acetyl derivatives of glucosamine and muramic acid, ribose or glycerol phosphates united by phospho-diester linkages, and other minor constituents like glucose, galactose, mannose, fucose or rhamnose. *Teichoic acids* are polymers of glycerol or ribitol phosphates combined with D-alanine and N-acetyl D-glucosamine units.

They make up 20-4 per cent of the dry weight of the cell wall. The *peptidoglycans* in *Staphylococcus* aureus include as major components the N-acetyl derivatives of glucosamine and *muramic acid* with horizontal strands of pentaglycine and vertical struts of tetrapeptides composed of L-alanine, D-glutamic acid, L-lysine and D-alanine residues. Both teichoic acids and peptidoglycans are antigenic determinants. The presence of D-amino acids in the feptidogl,can confers protection from attack by peptidases.

Peptide synthesis of this type does not involve the ribosomal system and the production of D-amino acids requires the presence of pyridoxal phosphate dependent racemases. The carbohydrate residues are united through the intervention of uridine nucleotides but as UDP is too polar to pass through the membrane, a carrier phospholipid functions in the transport system.

It is believed that penicillin blocks the synthesis of cell walls by interfering with the action of the glycopeptide transpeptidase concerned with the cross-linking reaction between the carbohydrate and amino acid residues. Some organisms can acquire resistance to the action of penicillin by producing an enzyme, penicillinase, which splits *penicillin* at the amide link of the betalactam ring to form an inactive compound, penicilloic acid. Such bacterial resistance is conferred by a plasmid with extra chromosomal DNA which can be transmitted to other bacteria to produce resistant strains. Cultures of both *sulphonamide* and *amethopterin*-inhibited bacteria accumulate in the medium 5-amino 4-imidazole carboxamide and its *ribotide* which are intermediates in the formation of nucleotides.

Other Heteropolysaccharides

Glycoproteins

These consist of *carbohydrate* linked to *a peptide* portion by *covalent* bonds only. *Mucopolysaccharides* may associate with peptides by *ionic* as well as by *covalent* bonds. Moreover, in contrast to mucopolysaccharides, glycoproteins contain no uronic acids only a few sulphate esters. The glycoproteins include some *plasma proteins, blood group substances* and *mucoproteins.*

Plasma Proteins include *lipoproteins* and *glvcoproteins*. Among the important plasma proteins are *albumin*, *thyroxine binding globulin*, *haptoglobin*, *fibrinogen* and the *immunoglobins*.

Blood group substances are located in the erythrocytes and are responsible for agglutination. The human types A, B and O yield amino acids and the sugars *D-galactose*, *glucosamine*, *galactosarnine* and *L-fucose* on hydrolysis.

Mucoproteins are characteristic of bone, connective tissue and salivary glands. *Galactosamine* and *glucosamine* are the sugars most commonly found in mucoproteins. They also contain *L-fucose* and *sialic acid.*

Murein

It is a *polysaccharide* linked to short *peptide chains.* It is found in the cell walls of many bacteria. It consists of alternating units of *N-acetylglucosamine* (NAG) and *N-acttylmuramic acid* (NAM) linked by its carboxyl group to the peptide chain consisting of four amino acids *(L-alanine*, *D-glutamic acid*, *L-lysine*, and *D-alanine*). Five *glycine* residues link each peptide chain to its neighbour. The enzyme *lysozyme* hydrolyses the chain into NAG-NAM disaccharides, resulting in the rupture of the cell wall.

Functions of Carbohydrates

The most important function of carbohydrates is to provide *energy* to the body. They are also *structural components* of tissues. Other functions include regulation of fat *metalism*, *protein sparing functions*, and function in the *digestive tract*.

Storage Substances of Potential Energy

Carbohydrates are storage substances of potential energy in animals. About 60% of the total energy requirement of man is provided by the breakdown of carbohydrates. One gram of carbohydrate on oxidation yields on an average four calories. Glucose supplies the immediate energy needed by tissues.

Glucose is the sole form of energy for the brain and other nervous tissue. Lack of glucose, or of oxygen for its metabolism, leads to rapid damage to the brain. Carbohydrate is stored in the body in the form of *glycogen*. About 100 gm of glycogen is stored in the liver and by its breakdown maintains the glucose level in the blood. Muscles contain about 200 to 240 gm glycogen. This glycogen is, however, utilised only by the muscles and is not available for regulating the blood sugar level.

Structural Component

Carbohydrates are important components of some structural materials of living organisms. *Monosaccharides are* important constituents of nucleic acids, coenzymes, flavoproteins and blood group substances, *Vitamin C* (ascorbic acid) is related to sugars *Immunopolysaccharides* play a part in the resi tance of infections. *Hyaluronic acid* is the viscous substance in the matrix of connective tissue.

Heparin prevents the clotting of blood. *Glucuronic acid,* which occurs in the liver, acts as a detoxifying agent by combining with toxic substances and bacterial by-products. *Chondroitin sulphates* are found in cornea, cartilage tendons, skin, heart valves and saliva. *Glycosides* are component of steroid hormones. *Galactolipins* are constituents of nervous tissue. The cell wall of plants and the polysaccharides in the capsule of bacteria are carbohydrates.

Regulation of Fat Metabolism

Some carbohydrates are essential for normal oxidation of fats. When carbohydrates are restricted in the diet there is more rapid metabolisation of fats. This results in the accumulation of incompletely oxidised intermediate products leading to *ketosis*. This is common in uncontrolled diabetes mellitus.

Protein-sparing Function

Carbohydrate is preferentially metabolised in the body as a source of energy as long as it is present in the required quantity. This spares protein for building of tissue. When there is deficiency of calories in the diet, however, fat and then protein are utilised for supplying energy.

Role of Gastrointestinal Function

Indigestible substances like cellulose, bemicellulose and pectins provide the bulk or roughage of food, and thus help the peristaltic movements of the digestive tract. Lactose promotes the growth of desirable bacteria in the small intestine.

These bacteria synthesize certain B-complex vitamins. Lactose also increases calcium absorption.

The demand for high-energy phosphate is not transient, and fuels must always be available to supply the processes generating high-energy phosphate. In order for an animal to interrupt its food intake—to eat meals rather than small snacks and to sleep for long intervals—it must store fuels at times of excess and be able to recover them at times of deprivation. These stores include a polymer of glucose: glycogen. (Plants use starch in a comparable way).

Glycogen appears in muscle cells as discrete particles with mass up to 2×10^7 daltons and more. Liver contains both individual particles and covalently linked aggregates as large as 10^9 daltons. The many hydroxyl groups make glycogen quite polar, and the particles contain 1.1 grams of bound water per grams of polysaccharide. Another 1 to 3 grams of water may interact with the particles. The particles also contain the enzymes responsible for the synthesis and later breakdown of glycogen.

The amount of glycogen varies widely, not only among different tissues but also within a tissue, depending upon the supply of glucose and the metabolic demand for energy. Most of the body's glycogen is in the liver and muscles. We shall mainly be concerned with these two tissues, but glycogen is also important to the economy of other tissues. In general, the content of glycogen in the liver varies greatly with the diet, while that in the muscles varies greatly with exercise.

Much of our knowledge about the glycogen concentration of human tissues comes from analysis by Swedish investigators of biopsy specimens from normal volunteers. (We are all indebted to university students and members of the fire department in Stockholm for contributing pieces of themselves of satisfy our curiosity). The data have some limitations, because only readily accessibly regions of the organs can be explored in living subjects, but the values are consistent with other quantitative information on human fuel metabolism. Results under various conditions from several laboratories are summarised.

A typical value for the liver glycogen content in a well-fed uman in near 400 millimoles of glucosyl residues (65 grams pry weight) per kilogram of tissue : there is less after fasting end more upon eating a lUigb.-carbohydrate meal.

The skeletal muscles typically have 85 millimoles of glucosyl readoes (14 grams) per kilogram of tissue, which does not change much with overnight fasting or with a highcarbohydrate diet. However, the content falls to 1 millimole per kilogram, or lower, after an hour or two of heavy work. After depletion in this way, high-carbohydrate diets on subsequent days can cause the level to rise as high as 300 milli-moles per kilogram.

Even though the liver usually has a higher concentration of glycogen than do the skeletal muscles, the muscles contain the bulk of the total

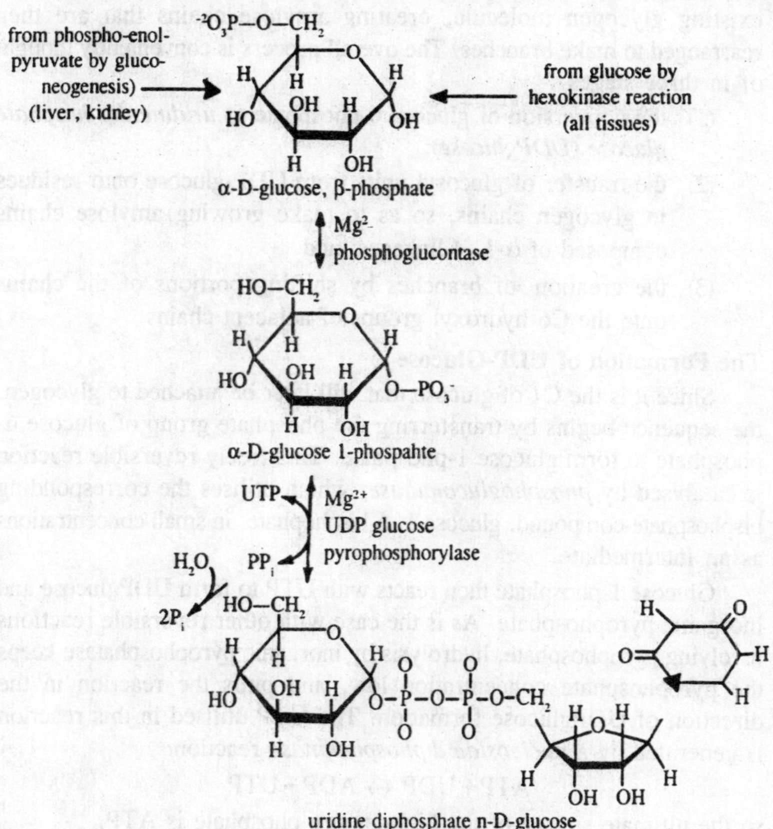

Figure 3.14: Storage of excess glucose residues as glycogen involves the formation of uridine diphosphare glucose (UDP-glucose).

glycogen store, owing to their large mass. A man weighing 70 kilograms will have about 28 kilograms of skeletal muscle, but only about 1.6 kilograms of liver. Given typical contents per kilogram of tissue, 400 millimoles in the liver and 85 millimoles in the muscles, the total store is 0.6 moles in the liver and 2.4 moles in the muscles. The total in the body, considering all organs, will therefore range somewhat over 3 moles in a well-fed individual and near 3 moles after an overnight fast (the postabsorptive state).

MECHANISM OF GLYCOGEN STORAGE

Glycogen is made by adding one glucosyl residue at a time to an existing glycogen molecule, creating amylose chains that are then rearranged to make branches. The overall process is conveniently thought of in three stages :

(1) the conversion of glucose 6-phosphate to *uridine dip hosphate glucose* (*UDPglucose*):

(2) the transfer of glucosyl units from UDP-glucose onto residues in glycogen chains, so as to make growing amylose chains composed of α-1, 4-linkages; and

(3) the creation of branches by shifting portions of the chains onto the C6 hydroxyl groups of adjacent chains.

The Formation of UDP-Glucose

Since it is the C1 of glucose that will later be attached to glycogen, the sequence begins by transferring the phosphate group of glucose 6-phosphate to form glucose 1-phosphate. This freely reversible reaction is catalysed by *phosphoglucomutase,* which utilises the corresponding bisphosphate compound, glucose 1, 6-bisphophate, in small concentrations as an intermediate.

Glucose 1-phosphate then reacts with UTP to form UDPglucose and inorganic pyrophosphate. As is the case with other reversible reactions involving pyrophosphate, hydrolysis by inorganic pyrophosphatase keeps the pyrophosphate concentration low, and pulls the reaction in the direction of UDPglucose formation. The UTP utilised in this reaction is generated by a *nucleoside diphosphokinase* reaction:

$$ATP + UDP \leftrightarrow ADP + UTP$$

so the ultimate source of the high-energy phosphate is ATP.

Formation of Amylose Chains

UDP-glucose donates glucose residues used for extending the terminal branches of glycogen in a reaction catalysed by *glycogen synthase*. The reaction is specific for the hydroxyl group on C4 of the glycogen residues, so the new glucosyl residue simply extends the 1, 4-chain. Since the nature of the chain is not changed by extension, the *glycogen synthase* reaction can be repeated indefinitely. A molecule of UDP is released for each glycosyl residue added.

Formation of Branches

If nothing else happened, the result of the glycogen synthase reaction would be long amylose chains composed only of 1, 4-bonds. However, cells storing glycogen also have a branching enzyme, *a glycosyl-4 : 6-transferase*, that catalyses the transfer of a segment of amylose chain onto the C6 hydroxyl of a neighbouring chain.

This enzyme moves a block of seven 1, 4-residues from a chain at least 11 residues in length and transfers it onto another segment of amylose chain at a point four residues removed from the nearest branch. (Transfer of seven residues is favoured, but specificity is not absolute, and this number is emphasized for clarity).

The new branch is therefore usually seven residues in length, and the remaining stub from which it was removed is at least four residues in length, but is more commonly six to nine residues. long. 1, 6-Glucosides have about 4,800 joules per mole less standard free energy content than do 1 ,4-glucosides, so the equilibrium of reaction favours branching.

Further Growth

The new branch and the remaining stub from which it was obtained can grow in length by addition of more 1, 4-glucosyl residues from UDP-glucose through the action of glycogen synthase. When enough have been added to extend them to at least 11 residues beyond the branch, a further transfer by the branching enzyme may create still another branch, which. also can grow in length.

What limits the size of the molecule and the number of particles is not known. There is little information to justify speculation, although large glycogen particles appear to be made of subunits of approximately the theoretical limit for tightly packed spheres. What is certain is that skeletal muscle does not accumulate glycogen indefinitely, even with a high concentration of blood glucose.

The content of glycogen in the liver can rise markedly under heavy dietary loading, with the formation of large aggregates, but even in this tissue glucose is absorbed slowly enough with more usual conditions so that an increased fraction of the excess supply is used for making fat after the glycogen content rises above 50 to 60 grams per kilogram of tissue. The mashed livers of force-fed Strasbourg geese, from which the pate de fore gras favoured by epicures is made, owe their special properties to high contents of both glycogen and fat.

Priming Glycogen Synthesis

Glycogen storage is mainly an extension of already existing polysaccharide chains, but there must be some way of making the core molecules that are to he enlarged. It is possible that reactions yet to be discovered put priming carbohydrate residues on a protein core. There is some evidence that this is not necessary, that the synthetic reactions used to enlarge glycogen will also make new starter molecules, although slowly.

UTILISATION OF GLYCOGEN AS FUEL

Phosphorolytic Cleavage

When glycogen is mobilised, all of the glucose residues appear as glucose 1-phosphate, except for one molecule of free glucose released from each 1,6 branch. The primary reaction is a simple phosphorolysis of 1, 4 bonds at the end of each branch. This reaction, which is catalysed by *phosphorylase*, shortens an outer branch by one residue, which appears as glucose 1-phosphate. The reaction is freely reversible, with the equilibrium ratio of P; to glucose 1-phosphate concentrations near 3.5. However, the concentration ratio in animal tissues is always greater than that, so the reaction always goes in the directions of formation of glucose 1-phosphate.

When a glucose residue is removed by phosphorlysis, the stortened 1,4-branch on the glycogen molecules can be attacked again to remove another residue, followed by further attack on the remainder. However, phosphorylase by itself will not catalyse the removal of all of the terminal residues. In anthropomorphic terms, it won't go near the branches.

Expressed more mechanistically, the specificity of the enzyme is such that the 1,4-bond attacked must be at least four residues removed from a 1,6-branch. Therefore rylase alone will not make the outer branches any shorter than four glucosyl residues. The function of this specificity is not clear. Glycogen completely sheared by phosphorylase to the four-residue stubs is called the phosphorylase limit dextrin, or ϕ-dextrin. (Dextrin is a term borrowed from starch chemistry, describing

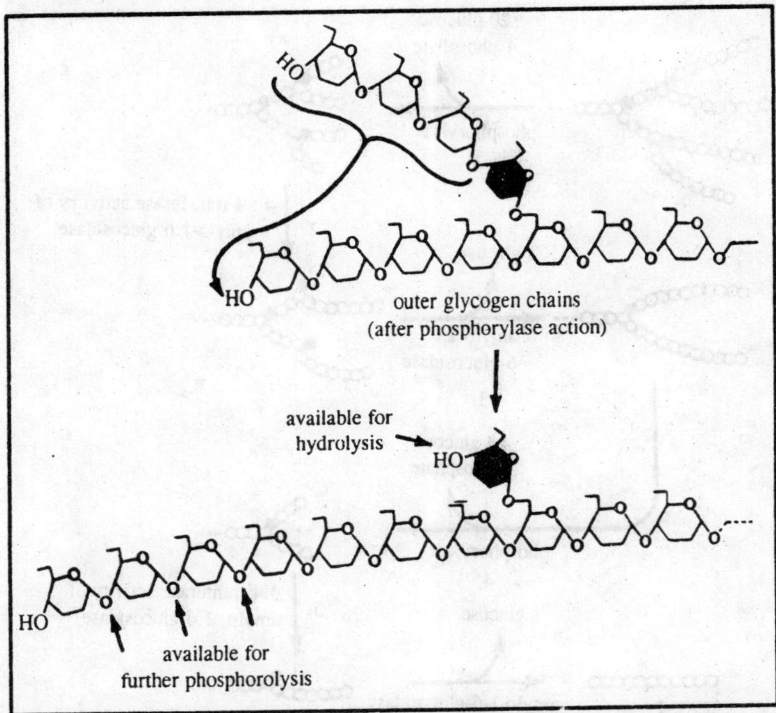

Figure 3.15: A transferase active site in amylo-1, 6-glucosidase the catalyses the transfer of three residues from the stubs of glycogen branches to the C4 hydroxyl groups of amylose stubs.

partially hydrolysed glucose homopolymers, and it is widely used for describing intermediates obtained during chemical investigation of structure, but this does not imply the occurrence of these discrete types of molecules in cells).

Removal of Branches

Further degradation of glycogen requires a second enzyme, *amylo-1,6-glucosidase*. This enzyme catalyses two successive reactions In the first reaction, it acts as a *glucosyl transferase* akin to the branching enzyme functioning during glycogen synthesis, but differing in that it transfers three glucosyl residues from a shortened branch onto another chain terminus so as to lengthen it.

Suppose that adjacent chains are shortened as much as possible at a branch by phosphorylase. Each will then be four glucosyl residues in length. The glucosyl 4 : 4-transferase activity in amylo-1,6-glucosidase catalyses the removal of three of the residues of the stub of the branch,

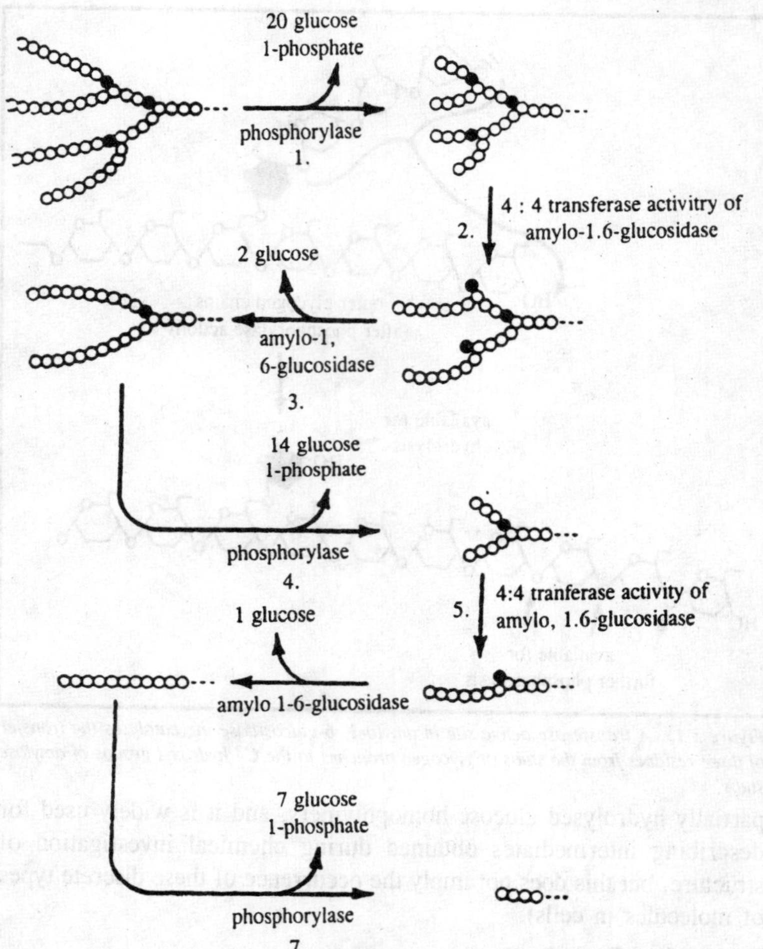

Figure 3.16: Summary of a sample sequence during glycogen breakdown. The diagram shows the fate of four outer chains and the next two tiers of branches from which they arise. Each chain is shown with nine residues beyond the outermost branch. The residues linked by 1,6 bonds and released as free glucose are indicated by solid circles.

leaving a single residue attached to C6. The enzyme transfers the three-residue package to the end of the other stub, which now has seven residues in 1,4-linkage and can therefore be attacked again by phosphorylase.

The amylo-1,6-glucosidase then catalyses its second reaction: hydrolysis of the single residue remaining attached by a 1,6-linkage at the branch to yield free glucose. The hydrolytic activity is absolutely specific for a 1,6-linkage to a single residue: this protects longer branches on glycogen from internally disruptive attack by the enzyme. Removal of the single branched residue exposes a straight chain of 1,4-linked residues down to the next branch, so phosphorylase can catalyse the removal of more residues as glucose- l-phosphate before reaching its limit of four residues next to a branch. What is the result ? The whole sequence causes a continual stripping of glucosyl residues as glucose 1-phosphate, except for the single residues at branches that are removed as free glucose. The degree of branching is such that 11 to 14 molecules of glucose 1-phosphate are formed for each molecules of free glucose released.

REGULATION OF GLYCOGEN STORAGE

The objectives of control over glycogen metabolism are to store glucose at times of plenty and mobilise the supply at times of need, while preventing the reaction of both storage and mobilisation from occurring at the same time. If both sets of reactions are proceeding simultaneously, the result is a futile cycle accomplishing nothing more than the hydrolysis of ATP. The critical reactions are these:

Synthesis

(1) glucose 1-phosphate + UTP $\rightarrow$ UDP-glucose + PP_i

(2) $PP_i + H_2O \rightarrow 2P_i$

(3) UDP-glucose $\rightarrow$ UDP + glycogen chain

(4) UDP + ATP $\rightarrow$ UTP + ADP

Mobilisation

(5) glycogen chain + $P_i \rightarrow$ glucose 1-phosphate

SUM ATP + $H_2O \rightarrow$ ADP + P_i

The strategy of control is to make glycogen phosphorylase and glycogen synthase respond to similar signals, but in opposite directions. When the glycogen phosphorylase becomes active, the synthase becomes inactive, and vice versa. In addition, *UDP-glucose* is made a competitive inhibitor of its own formation so that it will not accumulate when the synthase is inactive.

The nature of the signals regulating giycogen metabolism and the responses to them differ among tissues. In the muscles, the objectives is to mobilise glycogen as a fuel for contraction and to store it at rest if the glucose supply is adequate. Consequently, the signals are the intracellular concentration of Ca^{2+} and AMP, which rise when contraction begins, and the blood concentrations of adrenaline and insulin. Adrenaline signals a state of alert, perhaps requiring fuel expenditure, while insulin signals that glucose is available for storage.

The liver regulates glycogen supply so as to buffer the blood glucose concentration. After glucose is absorbed by the small bowel, it first goes to the liver *via* the portal vein and part is stored as glycogen. The liver can also store glucose residues synthesised by gluconeogenesis. When the blood glucose concentration falls, the liver breaks down glycogen and releases glucose into the blood to maintain the concentration.

Important signals for glycogen metabolism in the liver include the *glucose concentration* within hepatic cells and the blood concentrations of two pancreatic hormones, *insulin* to indicate an adequate glucose supply and *glucagon* to indicate a need for more glucose. In addition, the liver responds to stimulation of its *sympathetic nerve supply*, and to other hormones to be considered later.

Many of these signals act to regulate both glycogen phosphorylase and glycogen synthase. These enzymes are regulated in two ways: by combination with allosteric effectors and by covalent modification through phosphorylation and dephosphorylation.

CONTROL BY PHOSPHORYLATION

Glycogen synthase and glycogen phosphorylase each exists in active and inactive conformations. Allosteric effectors and phosphorylation regulate these enzymes by stabilising one conformation. Phosphate groups are added by transfer from, ATP in reaction catalysed by kinases, and they are removed by hydrolysis in reactions catalysed by phosphatases.

The kinases and phosphatases affecting the enzymes of glycogen metabolism also act on enzymes responsible for other metabolic processes that are discussed later. In general, *phosphorylation activates catabolic enzymes and inactivates anabolic* (*synthetic*) *enzymes*. This is true for the enzymes of glycogen metabolism. Phosphorylation of glycogen phosphorylase locks it in the active conformation. That is the P-form (phosphorylated form) is glycogen phosphorylase a ("*a*" for intrinsically active), but phosphorylation locks glycogen synthase in the inactive conformation, designated glycogen synthase *b*:

glycogen synthase *a* is *not* phosphorylated
glycogen synthase *b* is phosphorylated
glycogen phosphorylase *a* is phosphorylated
glycogen phosphorylase *b* is not phosphorylated.

The mechanisms for regulating the interconversion of active and inactive conformations of glycogen phosphorylase and glycogen synthase by the combination of allosteric effectors and phosphorylation-dephosphorylation reactions are considerably more complex than anyt^h^ing we have encountered before. Let us first dissect the responses in mammalian skeletal muscles.

REGULATION OF PHOSPHORYLASE IN MUSCLE

Phosphorylase is an enzyme with multiple binding sites on each of its two identical subunits. The enzyme clings to a glycogen particle like a barnacle on a rock, wrapping itself around a terminal amylose chain. This holding site has no catalytic function. Another crevice on the same concave face fits the terminal amylose chain to be cleaved with inorganic phosphate.

This catalytic site has a deeply buried molecule of *pyridoxal phosphate*. Pyridoxal phosphate is usually employed in reactions involving combination of its aldehyde group with amino groups on substrates, but here it is its phosphate group that is activated, probably to form a transient pyrophosphatelike intermediate, and no other example of this function is known.

The opposite face of glycogen phosphorylase, the convex face, is the control face. It has sites for binding AMP and for phosphorylation.

AMP Activates Glycogen Phosphorylase

Only the active conformation of the enzyme tightly binds AMP. Phosphorylase *a*, the P-form, is already in the active conformation, so combination with AMP does not affect its activity. Most of glycogen phosphorylase *b*, the dephospho form, is in the inactive conformation in the absence of AMP, but a rice in the concentration of AMP pulls the equilibrium toward the active conformation, even though the enzyme is not phosphorylated. Since a rising AMP concentration is a sensitive indicator of a fall in high-energy phosphate concentration, which will accelerate the Embden-Meyerhof pathway, its effect on glycogen phosphorylase is an important means of matching fuel supply and demand.

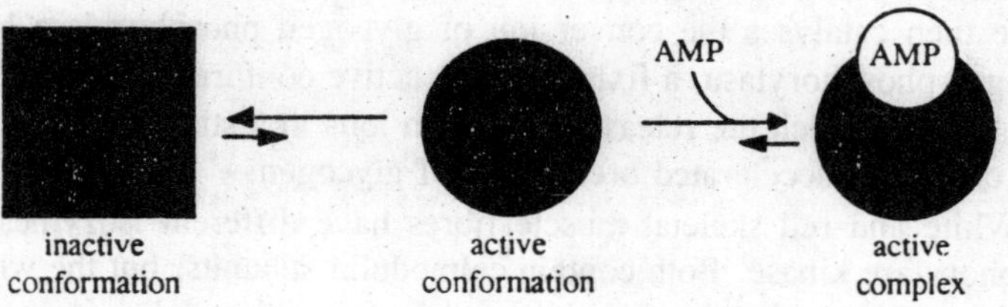

Calcium Ions Indirectly Activate Glycogen Phosphorylase

They affect the activity of another enzyme, phosphorylase kinase, which in turn catalyses the phosphorylation of glycogen phosphorylase:

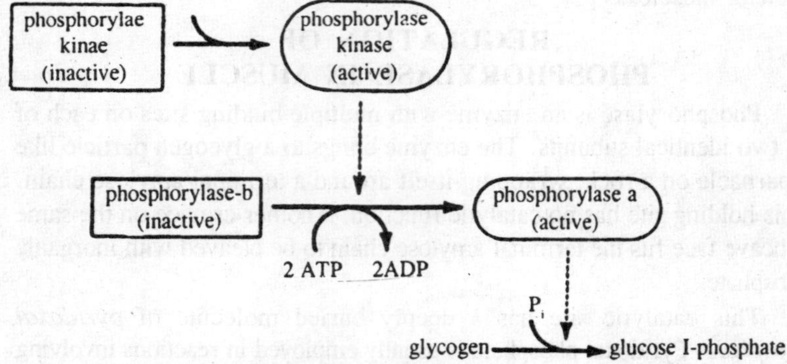

Phosphorylase kinase has the subunit composition $\alpha_4\beta_4\gamma_4\delta_4$. It is unusual in that the a subunit is *calmodulin, a* protein that also exists separately as an important regulatory protein in many kinds of cells.

Calmodulin binds for calcium ions per monomer, and its active conformation only appears when all the binding sites are filled.

The activation of calmodulin therefore depends upon the fourth power of the calcium ion concentration:

$$\text{calmodulin} + 4\text{Ca}^2 \rightarrow \text{calmodulin-Ca}_4 \text{ complex}$$

$$K = \frac{[\text{calmodulin-Ca}_4]}{[\text{calmodulin}][\text{Ca}^{2+}]^4}$$

This has the effect of creating a concentration threshold; only a small increase in $[Ca^{2+}]$ above the threshold is necessary to fully activate calmodulin.

When the calmodulin subunits within phosphorylase kinase are activated, the entire molecule shifts to an active conformation. The kinase then catalyses the conversion of glycogen phosphorylase b to glycogen phosphorylase, a fixing it in its active conformation, and this is the way in which the release of calcium ions in a stimulated muscle fiber causes an accelerated breakdown of glycogen.

White and red skeletal muscle fibres have different isozymes of phosphorylase kinase. Both contain calmodulin subunits, but the white muscle isozyme also binds additional active calmodulin from the sarcoplasm to further augment its catalytic activity. This makes the white fibers sensitive to lower [Ca^{2+}] than are red fibers; they mobilise glycogen more promptly upon stimulation, as is consonant with their role in rapid development of high tension for high power output.

When an animal is confronted with an emergency that may demand prompt and strenuous physical activity-flee or fightit is advantageous for it to mobilise glucose 1-phosphate rapidly from glycogen in its skeletal muscles so as to have the fuel available even before the signal for contraction is given. The advantage is achieved because emergencies of this sort are recognised in the central nervous system, which stimulates the adrenal medulla (the tissue in the core of the adrenal glands) to release adrenaline and the related noradenaline into the blood stream:

OH, OH; HO—C—H; CH_2—NH_3

coradrenaline

OH, OH; HO—C—H; CH_2—NH_2—CH_3

adrenaline

The human adrenal medulla releases more adrenaline than noradrenaline, but the reverse is true in some mammals. Both compounds are frequently lumped under the general name of *catecholamines*. They are frequently called *epinephrine* and *norepinephrine* in American medical circles because Adrenalin is a trademark. However, the tissue receptors for catecholamines are called *adrenergic receptors* by all.

Adenylate Cyclase and Cyclic AMP

Adrenaline stimulates the breakdown of glycogen in skeletal muscle by tripping a series of events that causes phosphorylase kinase to be locked in an active conformation by phosphorylation. These events begin with activation of an enzyme, *adenylate cyclase*, in the plasma membrane, of the muscle fibres.

This enzyme occurs as a complex spanning the plasma membrane, with the adrenaline receptors protein on the outside and the catalytic site on the inside. (Activation involves binding of GTP by a regulator protein in the complex.) Adenylate cyclase catalyses an intramolecular condensation of ATP to produce *adenosine* 3', 5'*-monophosphate*, more familiarly known as *cyclic AMP*.

The compound is sometimes designated as *3', 5'-AMP* and sometimes

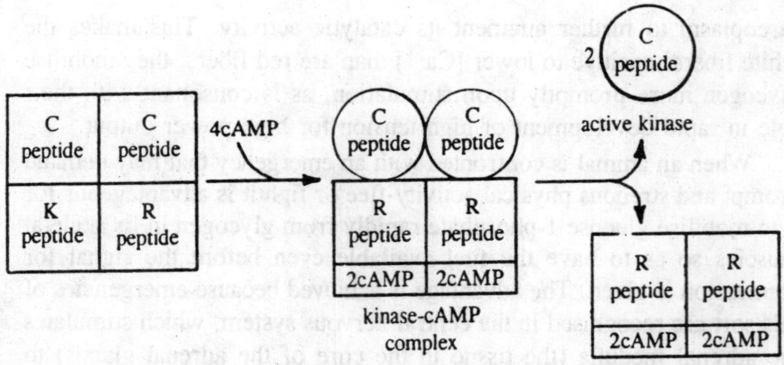

Figure 3.17: The activation of the CAMP-dependent protein kinase.

as *cAMP*. Cyclic AMP acts as a "*second messenger*", conveying the signal received in the form of adrenaline on the plasma membrane to the responsive enzymes within the cell.

Activation of cAMP-Dependent Protein Kinase

The major effect of cyclic AMP in eukaryotic cells is the activation of a particular protein kinase. The activatian occurs in this way : The protein kinase consists of two pairs of subunits: R_2C_2 in which R is a regulatory subunit and C is a catalytic subunit. The combined tetramer is inactive. Each of the regulatory subunits will combine with two molecules of cAMP, but when it does, the association with the catalytic subunits is weakened, and they are related as fully active monomers.

Effect of Protein Kinase on Phosphorylase Kinase

The cAMP-dependent protein kinase catalyses the transfer of phosphate from ATP to eight serine side chains in phosphorylase kinase. Each of the four g subunits has one site that is rapidly phosphorylated, stabilising the active conformation of the enzyme in the presence of Ca^{2+}. Each a subunit has one site that is less rapidly phosphorylated ; the function of this phosphorylation is discussed later.

The phosphorylation of phosphorylase kinase is the final link in a multienzyme cascade by which an elevation in blood adrenaline concentration causes conversion of glycogen to glucose 1-phosphate (and lesser amounts of glucose).

(1) Adrenaline is bound by a cell-surface receptor protein.

(2) A conformational shift occurs, permitting a neighbouring regulator protein to bind GTP.

(3) A further conformational shift exposes the catalytic site of a neighbouring adenylate cyclase protein to MgATP in the cytosol,

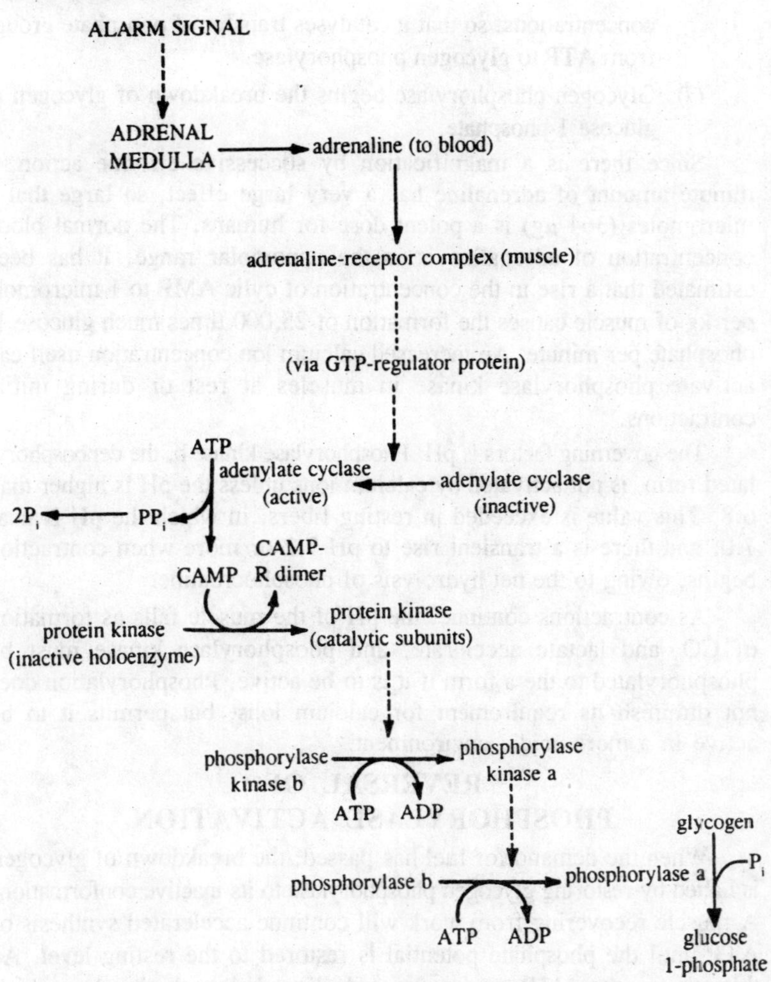

Figure 3.18 : A cascade of enzyme act rations translates the appearance of adrenaline into an accelerated formation of glucose I-phosphate from glycogen.

and cyclic AMP is formed.

(4) Cyclic AMP combines with a particular protein kinase complex in the cytosol, causing it to release its catalytic subunits.

(5) The catalytic subunits of the protein kinase react readily with the β subunits, and less readily with the α subunits, of phosphorylase kinase on glycogen particles.

(6) Phosphorylation of the β subunits of phosphorylase kinase fixes the enzyme in its active conformation, even at low calcium ion

concentrations, so that it catalyses transfer of phosphate groups from ATP to glycogen phosphorylase.

(7) Glycogen phosphorylase begins the breakdown of glycogen to glucose 1-phosphate.

Since there is a magnification by successive enzyme action, a minute amount of adrenaline has a very large effect, so large that 2 micromoles (364 μg) is a potent dose for humans. The normal blood concentration of adrenaline is in the nanomolar range. It has been estimated that a rise in the concentration of cylic AMP to 1 micromole per kg of muscle causes the formation of 25,000 times much glucose 1-phosphate per minute. An increased calcium ion concentration itself can activate phosphorylase kinase in muscles at rest or during initial contractions.

The governing factors is pH. Phosphorylase kinase b, the depbosphorylated form, is not activated by calcium ions unless the pH is higher than 6.8. This value is exceeded in resting fibers, in which the pH is near 7.0, and there is a transient rise to pH 7.2 or more when contraction begins, owing to the net hydrolysis of phosphocreatine.

As contractions continue, the pH of the muscle falls as formation of CO_2 and lactate accelerate, and phosphorylase kinase must be phosphorylated to the a form if it is to be active. Phosphorylation does not diminish its requirement for calcium ions, but permits it to be active in a more acidic environment.

REVERSAL OF PHOSPHORYLASE ACTIVATION

When the demand for fuel has passed, the breakdown of glycogen is halted by restoring glycogen phosphorylase to its inactive conformation. A muscle recovering from work will continue accelerated synthesis of ATP until the phosphate potential is restored to the resting level. As this occurs, the AMP concentration declines below the level at which AMP can fix glycogen phosphorylase *b* in its active conformation. However, glycogen phosphorylase a remains active in the absence of AMP, and several additional events occur in a resting muscle to convert glycogen phosphorylase a back to glycogen phosphorylase b and prevent its reactivation in the absence of further excitation of the muscle. These events include

(1) Activation of protein phosphatases to hydrolyse the phosphate groups from glycogen phosphorylase *a* and phosphorylase kinase *a*;

(2) Inactivation of the cAMP-dependent protein kinase through hydrolysis of cyclic AMP by cyclic nucleotide phosphodiesterases:

(3) Cessation of further formation of cyclic AMP, after the adrenaline concentration falls, through prompt hydrolysis of GTP by the regulatory protein in the adenylate cyclase complex.

(4) Rapid destruction of circulating adrenaline to lower its concentration in the absence of continued stimulation of the adrenal medulla.

The activation of protein phosphatases also involves the enzyme cascade. Protein phosphatase-1 is an enzyme capable of hydrolysing phosphate groups on many proteins. It occurs in many tissues, including the muscles, where it will convert glycogen phosphorylase *a* to glycogen phosphorylase *b*, and also will remove phosphate groups from the β subunits of phosphorylase kinase, provided that the α subunits are also phosphorylated. This phosphatase is regulated by the phosphorylated form of another protein known as protein inhibitor-1.

Protein inhibitor-1 is phosphorylated by the same cAMPdependent protein kinase that converts phosphorylase kinase *b* to phosphorylase kinase *a*. Using these facts, we can expand the events in the cascade regulation of glycogen breakdown:

(1) The cAMP-dependent protein kinase rapidly phosphoryiot only the β-subunits of phosphorylase kinase, making it active, but also protein inhibitor-1, which combines with protein phosphatase-1 to inactive it.

(2) The protein kinase slowly phosphorylates the a-subunits of phosphorylase kinase, making it susceptible to attack by active protein phosphatase. During the interval between phosphorylation of the β- and α-subunits, phosphorylase kinase will remain fully active, ensuring some minimum response to the adrenaline signal.

(3) Protein phosphatase-1 can catalyse the removal of the phosphate group from its own inhibitor, and it begins to do so, causing the inhibitor to dissociate.

(4) As the active protein phosphatase is released, it removes the phosphate groups from the β-subunits of phosphorylase kinase, Making it inactive.

(5) The active protein phosphatase-1 will not attack glycogen phosphorylase a that has AMP bound to it. If the AMP

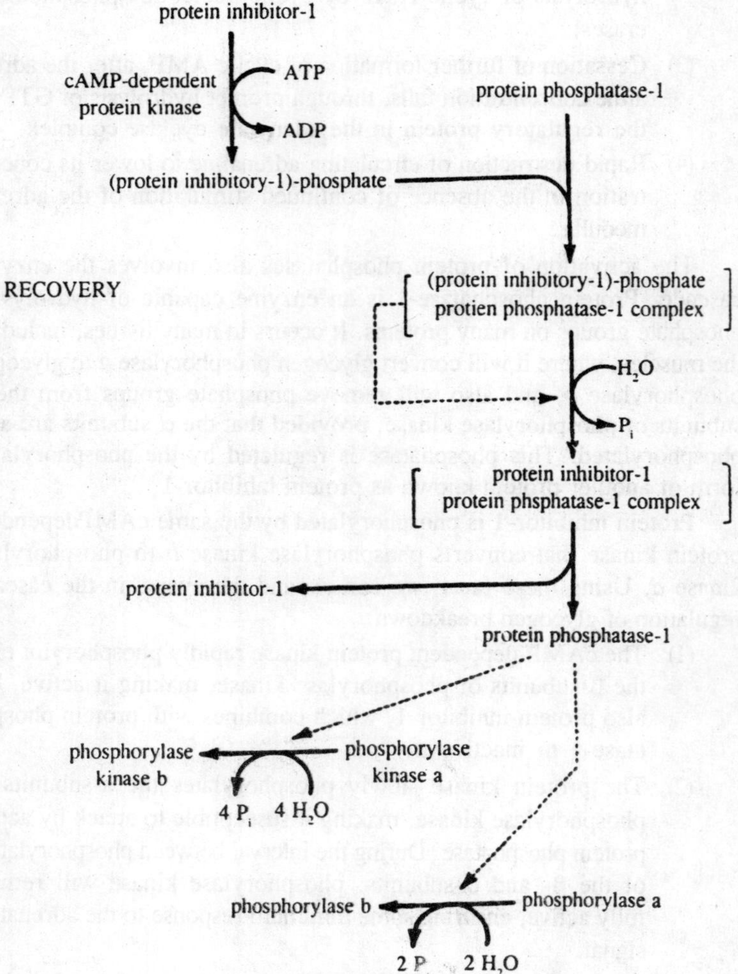

Figure : 3.19 : Regulation of protein phosphatase-1, the principal enzyme inactivating phosphorylase.

concentration is low, then the protein phosphatase will convert glycogen phosphorylase a to glycogen phosphorylase *b*, which is inactive at low AMP concentrations. A high AMP concentration thus not only activates glycogen phosphorylase *b*, it also maintains phosphorylase a in its inherently active phosphorylated state, once it is formed.

(6) A protein phosphatase-2 is also present to catalyse the ,moval of the phosphate groups from the α-subunits of phoshorylase kinase, returning the enzyme to its resting state, ready to go through the activation cycle again.

REGULATION OF GLYCOGEN SYNTHASE IN MUSCLE

Glycogen synthase also has active and inactive conforma ons. When the enzyme is not phosphorylated, it is mostly in ie active conformation. Phosphorylation stabilises the inactive information of this enzyme. Each of the identical subunits in lycogen synthase may be phosphorylated by three kinds of inases, each acting at different locations on the polypeptide: hain. Phosphorylase kinase is the most effective; only one site need be phosphorylated by it to stabilise the inactive conformation (glycogen syntbase *b).*

A Ca^{2+}-dependent (cAMPindependent) protein kinase is next most effective, and it will add as many as four phosphate groups to each monomer. The cAMP-dependent protein kinase is least effective ; it pbosphorylates two sites in each monomer, but without generating full syntbase activity. The inactivation of glycogen synthase in muscle is therefore stimulated by the same events that activate glycogen phosphorylase; a rise in the concentrations of adrenaline in the blood, or calcium in the sarcoplasm, or both.

Regulation by Glucose 6-Phosphate. Glucose 6-phosphate activates glycogen syntbase *b,* raising the V.., and lowering the *KM* for UDP-glucose. It seems likely that this effect would be useful for gathering up glucose 6-phosphate and storing the glucosyl group as glycogen at any time an accumulation of the compound isn't being utilised for energy production. Whether this is true for mammalian muscle is disputed by many, but it so happens that frogs have a glycogen synthase with an absolute requirement for glucose 6-phosphate for activity, even in the free from, and it is therefore difficult to doubt the physiological relevance of the effect to the croakers.

Regulation by Glycogen Concentration. Glycogen inhibits the removal of phosphate from glycogen syntbase ***b*** by a protein phosphatase. As glycogen accumulates in the muscle, it therefore shuts off its own production by decreasing the conversion of glycogen synthase ***b*** to its more active free form. This effect is the most likely explanation of why animals on high carbohydrate diets don't store massive quantities of glycogen in their muscles.

PHOSPHORYLATION OF MYOSIN

Phosphorylation and dephosphorylation are also used to control the contractile apparatus, at least in smooth muscles. The light chains of myosin are partially phospborylated in all muscles by a specific protein kinase requiring the calciumcalmodulin complex for activity. The physiological function of the phosphorylation in striated muscles is unknown, but it is absolutely necessary for contraction of smooth muscles, in which it provides a locking device to sustain the contracted state without additional expenditure of ATP. There is a corresponding protein phosphatase and inhibitor protein to control the removal of the phosphate groups.

GLYCOGEN METABOLISM IN THE LIVER

The turnover of glycogen in the liver is regulated through the same kinds of mechanism seen in striated muscles. However, several of the enzymes involved are isozymes of those occurring in muscles and are regulated by metabolites in different ways. One of the important modulations of enzyme activity in the liver is a combination of glucose with phosphorylase a, which inactivates the enzyme. This means that an elevated concentration of glucose inhibits the breakdown of liver glycogen to form more glucose. In addition, the glucoseglycogen phosphorylase a complex is a better substrate for protein phosphatase-1 than is glycogen phosphorylase a alone, so the presence of glucose accelerates the conversion of glycogen phosphorylase a to the less active *b* form.

Glucagon and Insulin

Secretion by Pancreas

The pancreas of higher animals is both an exocrine organ that elaborates enzymes to digest food and an endocrine organ that secretes hormones into the blood. The pancreas contains organised nests of cells known as the *islets of Langerhans*, which play a major role in regulating metabolism by secreting insulin, and glucagon. We have already encountered insulin, which consists of two polypeptide chains linked by disulfide bonds, as an example of modification of a polypeptide after its synthesis the cells forming insulin -ire known as *beta cells*. Glucagon is also a polypeptide, but has only some 29 residues in the known examples, and it is secreted by different cells in the islets, the *alpha cells*. (More is said about pancreatic endocrine function).

These hormones have opposing effects on glucose metabolism. The alpha cells are stimulated to release glucagon when the blood glucose concentration falls, and the beta cells release .insulin when the glucose concentration rises.

The regulation of glycogen metabolism in the liver illustrates a general principle ; *Adenylate cyclase is activatea by different hormones in different tissues. Adenylate cyclase in the liver is complexed with receptors for glucagon, whereas it is not in muscles. Therefore, glucagon secretion accelerates the breakdown of glycogen in liver* through activation of the cAMPdependent protein kinase, but it has no direct effect on glycogen metabolism in the muscles. When glycogen is being broken down in the liver, most of the resultant glucose 6-phosphate is hydrolysed to glucose by the action of glucose-6-phosphatase. The glucose moves out of the liver into the blood and is carried to other tissues.

On the other band, insulin promotes the storage of glucose as glycogen in both the liver and the muscles. It is now over 70 years since a graduate student at the University of Chicago, E.L. Scott, demonstrated that the pancreas secreted a factor lowering the blood glucose concentration, and 60 years since an insulin preparation was first used to save a diabetic patient, but we still do not know the details of bow insulin works.

We know that it causes more glucose transporter proteins to be moved to the plasma membrane of cells in at least some sensitive tissues, but this does not explain its internal actions on cellular metabolism. Relief may be *in* sight. There is evidence that combination of insulin with its receptors causes the release of a small protein from the plasma membrane into thekytoplasm, where it acts as a second messenger. This messenger is not derived from insulin directly ; it may be a fragment of a larger protein in the plasma membrane.

It is belived to have three effects : an inhibition of the cAMP-dependent protein kinase, an activation of a protein phosphatase-2 that attacks glycogen synthase *b* (not the one attacking glycogen phosphorylase *a*), and an activation of the protein phosphatase component of the mitochondrial pyruvate dehydrogenase complex. Therefore, insulin causes glycogen synthase and pyruvate dehydrogenase to become more active and suppresses activation of glycogen phosphorylase.

The result is increased synthesis of glycogen in the liver and muscles, and an increased conversion of pyruvate to acetyl coenzyme A in these organs. These effects are in addition to the action of insulin in accelerating glucose transort into muscles.

In general, insulin secretion rises when the secretion of glucagon falls, and *vice versa*. Suppose both are being secreted. There is some protection against simultaneous activation of the the synthase and glycogen phosphorylase because glycogen phosphorylase *a*, but not the glucose-glycogen phosphorylase a complex, is a potent inhibitor of the action of protein phosphatase-2 on glycogen synthase *b*. Glucagon seemingly can override the effects of insulin on glycogen turnover in the liver, provided that the glucose concentration is low.

SUMMARY OF GLUCOSE HOMEOSTASIS

Let us pull all of this together by recapitulating the events that occur when the blood glucose concentration rises and falls.

Excess Glucose

We eat and the blood glucose concentration rises. There are three important direct results

1. The pancreatic islets put out less glucagon and more insulin. Less glucagon means less cyclic AMP formed in the liver, and less active protein kinase. With less active protein kinase, there is less formation of active glycogen phosphorylase and less inactivation of glycogen synthase. More insulin also results in more active glycogen synthase in both the liver and the muscles and more rapid transport of glucose into the muscles.
2. Glucose combines with glycogen phosphorylase in the liver. This causes increased loss of active glycogen phosphorylase:
3. Increased blood glucose concentration results in the net formation of glucose 6-phosphate in the liver, since the liver glucokinase is not saturated with substrate and therefore can respond to changes in blood glucose concentration.

All of these things taken together mean an elevation in blood glucose concentration is itself a sufficient signal to cause further storage of glucose as glycogen in both the liver and muscles. The increased glycogen synthase activity in the muscles causes increased removal of glucose 6-phosphate glucose 6-P↔glucose 1-P→UDP-glucose-3glycogen. Since glucose 6-phosphate is an inhibitor of its own formation by the hexokinase reaction, its increased removal to form glycogen also promotes increased uptake of glucose in the muscle.

The deposition of glycogen in the muscles will proceed only until a characteristic level is reached even if elevated concentrations of glucose persist. However, the liver will continue to store more glycogen, although at an increasingly slower rate. We shall see in the following chapters that more and more glucose is converted to fat for storage as the glycogen reserves begin to be filled.

Deprivation of Glucose

The blood concentration falls as more time elapses from the preceding meal. Here we have the opposite effects

1. The pancreatic islets put out more glucagon and less insulin. More glucagon means more cyclic AMP formed in the liver and a more active protein kinase. The protein kinase causes the conversion of glycogen synthase to its inactive phosphorylated form, the conversion of glycogen phosphorylase to its active phosphorylated form, and the phosphorylation of protein inhibitor-1, which inactivates protein phosphatase-1.
2. The fall in the amount of glucose slows the hydrolysis of glycogen phosphorylase a because phosphorylase a without glucose is a poorer substrate for protein phosphatase-l than is its glucose complex.
3. The low concentration of glucose results in a low rate of the liver glucokinase reaction so that the liver adds glucose to the blood rather than removing it.

What are the results ? The changes in enzyme activity in the liver stop the formation of glycogen and accelerate its degradation. The resultant rise in the concentration of glucose 6-phosphate will accelerate the hydrolysis of this compound to form free glucose, which will diffuse into the blood to prevent a further drop in concentration.

The maintenance of the blood glucose concentration will enable the brain to continue using this fuel at its normal rate, and the supply for this purpose is conserved by the simultaneous slowing of glucose uptake and of glycogen formation in the muscles, owing to the declining insulin levels.

THE EFFICIENCY OF GLYCOGEN STORAGE

What is the price the organism pays for storage of excess glucose as glycogen ? We can show it is only 3 per cent of the potential for generating ATP. Let us put the question in this way: How much ATP will be generated per glucosyl group that disappears during combustion if the glucose has been stored as glycogen, compared to the amount generated when glucose is used directly without storage ?

It requires two high-energy phosphates to store one glucose residue as glycogen according to the balance:

glucose + ATP → glucose 6-phospbate + ADP

glucose 6-phosphate → glucose 1-phosphate

glucose 1-phosphate + UTP → UDP-glucose + PP_i

UDP-glucose + glycogen → glucosyl-glycogen + UDP

$PP_i + H_2O \rightarrow 2P_i$

UDP + ATP → UTP + ADP

SUM : glucose + 2ATP + glycogen + H_2O →

glucosyl-glycogen + 2(ADP + P_i)

Where is the high-energy phosphate to be obtained ? It may be from combustion of any fuel, but we can approximate the net cost by assuming that glucose is the only fuel available, and that part of its supply must be burned immediately in order to store the remainder, and the complete combustion of one glucose molecule will generate 37.5 ATP in white fast-twitch fibers. From this, we can estimate that 5.3 per cent of the gluose will be burned to store the remaining 94.7 per cent as glycogen

947 × 2ATP consumed ≅ 053 × 35.5 ATP produced

When the stored glycogen is later used, approximately 95 per cent of it will be converted to glucose 1-phosphate; the remaining 5 per cent will be released as free glucose from hydrolysis of the 1,6 branches.

Suppose that there were 100 millimoles of glucose originally available. 94.7 millimoles were stored as glycogen, while 5.3 millimoles were burned to provide the energy for storage. When the 94.7 stored millimoles are recovered, 5 per cent, or 4.7 millimoles, is released as free glucose, and 90 millimoles are obtained as glucose 1-phosphate. Oxidation of glucose 1-phosphate generates 36.5 ATP ; therefore 90 × 36.5 = 3,285 millimoles of ATP will be formed.

Out of the original 100 millimoles of glucose, 4.7 millimoles reappeared as glucose during the breakdown of glycogen ; therefore, only 95.3 have disappeared to generate 3,285 millimoles of ATP, for a yield of 34.5 ATP per glucose molecule disappearing. This compares with 35.5 that could be obtained by direct oxidation of glucose. The cost of glycogen storage is therefore only 3 per cent of the potential ATP—a small price for an immediately available supply of rapidly combustible fuel. (This analysis neglects the cost of carrying. the extra weight of the stored fuel.)

Efficiency of the Cori Cycle

A similar analysis can be made for the Cori cycle, in which part of the lactate made from glycogen in white muscle fibers is converted to glucose in the liver and returned to the muscles for re-storage as glycogen. This calculation is again a simplification of complex balances in which it is assumed that combustion of lactate by the liver provides the high-energy phosphate for gluconogenesis. It goes like this:

Skeletal muscles: 100 millimoles of glucose residues in glycogen yield 200 millimoles of lactate plus 300 millimoles of ATP.

Liver: 200 millimoles of lactate arrive in the blood from the muscles.

1. 31-3 millimoles are oxidised to CO_2 to generate 548 millimoles of ATP (17.5 millimoles per lactate, utilising the malate-aspartate electron shuttle).
2. The remaining 168.7 millimoles of lactate are convert-ed to 84.4 millimoles of glucose, utilising the 548 millimoles of ATP generated in step 1.

Skeletal muscles: 84.4 millimoles of glucose arrive in the blood from the liver.

1. 94.7 per cent of the glucose or 79.9 millimoles can be stored as glycogen by burning the other 5.3 per cent to provide the energy.
2. This replaces all but 20.1 millimoles of the original glucose residues used in the muscle; therefore, the muscle gained the 300 millimoles of ATP used initially for contraction at the expense of 20.1 millimoles of glucose, for a yield of 14.9 ATP per glucose residue consumed.

This value is 40.9 per cent of the 36.5 ATP generated by total oxidation in the muscle of a glucose residue in glycogen.

GENETIC DEFECTS IN GLYCOGEN METABOLISM

A number of humans have been discovered who are deficient in one of the enzyme activities necessary for glycogen metabolism. The combined incidence of these defects is about 1 in 40,000 births. One case might arise per year in Los Angeles. However, they add to the catalogue of genetic defects that does demand the attention of every pediatrician. In addition, their effects have promoted our understanding of normal carbohydrate metabolism.

A defect may occur in the formation of any of the proteins necessary for either the breakdown or the synthesis of glycogen, so that the result could be either a high or a low content of glycogen with a normal structure, or the presence of glycogen with abnormalities in the degree of branching. A classification of the resultant clinical conditions is given in Table 3.1. Hewever, it ought to be remembered that genetic errors can affect the formation of a protein in various ways; the rate of synthesis of normal protein may be changed, or mutations of coding may affect amino acid composition at any point in the molecule. The result is that there may be many types of deficiency of a given enzyme, some more severe in their consequences than others, so the rigid classification of the table does not adequately convey the actual spectrum of diseases.

A genetic defect may affect all tissues with an active glycogen metobolism. However, in some cases, a defective gene will disturb glycogen metabolism in only a few tissues because normal genes for different isozymes are still being expressed in other organs. Thus, defects in glycogen phosphorylase may be peculiar to the liver or to the skeletal muscles, because different isozymes are made in these organs.

Two decades ago it was believed that glycogen is synthesized by glycogen phosphorylase, because its reaction is readily reversed in the test tube. A young man was discovered who was weak and who developed severe pain in his muscles after modest exercise, and he was shown to have high concentrations of glycogen and an absence of glycogen phosphorylase in his muscles.

This proved that he was not synthesizing glycogen by the phosphorylase reaction and provided strong supporting evidence for the then newly discovered UDP-glucose pathway. (The condition is known as *McArdle's disease*, after its discoverer. A psychiatrist engagingly confesses that he thought McArdle's patient was displaying classic hysteria owing to an unhappy childhood. We are not told how much the phosphorylase deficiency may have contributed to the unhappy childhood).

McArdle's disease tells us something else. Impairment of lycogen utilisation in skeletal muscles is not fatal. The impairment is very real ; there isn't any unsuspected route for utilising the polysaccharide, and this is easily demonstrated by shutting off the blood supply to an arm with a tourniquet. Clenching of the fist causes a sharp rise in lactate concentration in a normal individual, ending in painful tetany, but there is no rise in a patient with McArdle's disease.

However, the importance of carbohydrate metabolism for full efficiency is shown by the weakness of the patient and is supported by the uncommon occurrence of the genetic lesion in the populationthis is a mutation that is eliminated rapidly.

This view of glycogen metabolism as a convenience for full efficiency

Table 3.1 : Glycogen storage diseases

Deficient Enzyme	*Name*	*Type*	*Site of Deficiency*	*Incidence*
glucose 6-phosphatase	von Gierke's disease	I	liver, kidneys	1 : 200,000
lysosomal a-glucosidase	Pompe's disease	II	all tissues	1 : 200000
amylo-1, 6-glucosidase	limit dextrinosis	III	all tissues	1 : 200000
branching enzyme	amylopectinosis	IV	all tissues	very low
glycogen phosphorylase	McArdls's disease	V	skeletal muscle	low
glycogen Phosphorylase	Here' disease	VI	liver	?
phosphofructokinase		VII	skeletal muscle, erythrocytes	very low
adenyl cyclase (?)		VIII	brain, liver	very low
phosphorylase kinase		IX	liver, other tissues (not muscle)	1 ; 100,000
cAMP-sensitive otein kinase		x	liver, muscles	very low

Note : Type numbers VI, VIII, and IX have been assigned to different deficiencies by some authors.

of muscles rather than as an absolute necessity is reinforced by the discovery of individuals with a deficiency of phosphofructo kinase in skeletal muscles, who have the same clinical picture found in McArdle's disease. They not only accumulate glycogen because of an inability to use it, they can't even use glucose taken up from the blood stream directly and they still survive.

Accumulation of normal glycogen in the liver causes massive enlargements of the organ, so much so that it may occupy a large fraction of the abdomen in affected children. This in itself causes surprisingly little difficulty; infants without glycogen phosphorylase in the liver grow to be adults. However, if the failure is due to an absence of glucose-6phosphatase, which prevents the use of stored glycogen to maintain the blood glucose concentration between meals, the consequences without treatment are grave.

The blood glucose concentration falls, while the lactate concentration rises. The brain is usually damaged, perhaps owing in part to the absence of glucose as a fuel and in part to the lactic acidosis. The treatment is repeated feeding of carbohydrate at two to threehour intervals night and day. If the infant can be brought through the first four years, his chances for gradual adjustment to a less heroic feeding schedule are good.

Defects in phosphorylase kinase are interesting because two patterns of inheritance are known. One is an autosomal recessive and the other is an X-linked recessive. If we bad only this information, we could infer that the kinase contains at least two polypeptide chains coded by genes in different chromosomes, and we have already seen that the kinase contains some three different chains.

In addition, another kind of phosphorylase kinase deficiency has been discovered that affects both the liver and the muscles, indicating that these isozymes have some common subunit. One of the most damaging storage diseased results from accumulation of glycogen by lysosomes. The enzyme affected is an ac-glucosidase that attacks either 1,4 or 1,6 linkages and is not involved in the major routes of glycogen metabolism.

It probably is a device for removing glycogen trapped during the normal scavenging function of lysosomes, which become filled to the bursting point when the enzyme is missing. All tissues are affected, but damage to the heart is usually the immediate cause of death, which occurs in infancy.

4

LIPIDS

Lipids that yield salts of fatty acids (or "*soaps*") upon hydrolysis are called *Saponifiable*. The alcohol components c such compounds may be glycerol, glycerol derivatives, or long chain alcohols. If lipids do not contain fatty acids in ester o amide linkage, they are usually categorised as *nonsaponifiable*. This group includes the *terpenes*, steroids, and the *prostaglandins*.

FATTY ACIDS AND THE PROSTAGLANDINS

Fatty acid is the name given to any acid comprised of a hydrocarbon chain with a terminal carboxyl group. Free fatty acids are found in only negligible amounts in *mammalian cells* and tissues. They occur principally in ester form. The most abundant have an even number of carbons, ranging from 14 to 22.

The structures of the 18 fatty acids found in significant quantities in mammalian lipids are listed in Table elsewhere in this chapter. The acid strengths of the saturated fatty acids are not influenced significantly by the length of the hydrocarbon chain.

The pK′ values of the carboxyl groups are approximately 4.8, comparable with that of acetic acid. On the other hand, melting point and solubility in water are markedly affected by the number of carbons. For example, the solubility of butyric acid is 5.6 per cent, whereas that of caproic acid is only about 0.4 percent. The melting points are also influenced by chain length, as seen in Table elsewhere in this chapter.

The presence of a carbon-carbon double bond makes a fatty acid distinctly different from its fully saturated counter-part. The melting

Table 4.1: Fatty Acids of Mammalian Tissues.

Formula	*Systematic Name*	*Common Name*	*Symbol*	*Melting Point (°C)*
Saturated fatty acids				
$CH_3(CH_2)_2COOH$	n-Butanoic	Butyric	4:0	
$CH_3(CH_2)_4COOH$	n-Hexanoic	Caproic	6:0	
$CH_3(CH_2)_3COOH$	n-Decanoic	Capric	10:0	
$CH_3(CH_2)_{10}COOH$	n-Dodecanoic	Lauric	12:0	44.2
$CH_3(CH_2)_{12}COOH$	n-Tetradecanoic	Myristic	14:0	53.9
$CH_3(CH_2)_{14}COOH$	n-Hexadecanoic	Palmite	16:0	63.1
$CH_3(CH_2)_{16}COOH$	n-Octadecanoic	Stearic	18:0	69.6
$CH_3(CH_2)_{18}COOH$	n-Eicosanoic	Arachidic	20:0	76.5
$CH_3(CH_2)_{22}COOH$	n-Tetracosanoic	Lignoceric	24:0	86.0
Unsaturated fatty acids				
$CH_3(CH_2)_5CH=CH(CH_2)_7COOH$	Hexadec-9-enoic	Palmitoleic	$16:1^{\Delta 9}$	—0.5
$CH_3(CH_2)_7CH=CH(CH_2)_7COOH$	Octadec-9-enoic	Oleic	$18:1^{\Delta 9}$	13.4 y
$CH_3(CH_2)_4CH=CHCH_2CH$ $=CH(CH_2)_7COOH$	Octadeca-9,12-dienoic	Linoleic	$18:2^{\Delta 9,12}$	-5

(Contd.)

Table 4.1 (Contd.)

$CH_3CH_2CH{=}CHCH_2CH{=}CHCH_2CH{=}CH(CH_2)_7COOH$	Octadeca-9,12, 15-trienoic	Linolenic	$18:3^{\Delta 9,12,15}$	–11
$CH_3(CH_2)_4(CH{=}CHCH_2)_3{=}CH(CH_2)_3COOH$	Eicosa-5,8,1 1, 14-tetraenoic	Arachidonic	$20:4^{\Delta 5,8,11,14}$	–49.5
$CH_3(CH_2)_9CH{=}CH(CH_2)_{12}COOH$	Docos-l3-enoic	Erucic	$24:1^{\Delta 13}$	
$CH_3(CH_2)_7CH{=}CH(CH_3)_{14}COOH$	Tetracos-15-enoic	Nervonic	$24:1^{\Delta 15}$	
Hydroxy fatty acids				
$CH_3(CH_2)_{21}{-}C(OH)(H){-}COOH$	α-OH-tetracosanic	Cerebronic	24:α-OH	
$CH_3(CH_2)_7CH{=}CH(CH_2)_{13}{-}C(OH)(H){-}COOH$	α-OH-tetracos-15-enoic	a-OH-nervonic	$24;\alpha\text{-}OH^{1\Delta 15}$	

points of the unsaturated acids are lower and also their solubility in a solvent such as ethanol is greater. This difference in physical properties of saturated and un-saturated lipids is due in part to the conformational peculiarities of a hydrocarbon containing carbon-carbon double bonds. Because a C=C bond cannot be rotated, the substituents on these carbon atoms may be either proximal (cis) or distal (trans) with respect to their spatial orientation.

The chain of a fatty acid of the *cis* configur-ation can have a kind produced by the C=C, resulting in a conformation distinctly different from the transform. The *predominant* form of unsaturated fatty acids in nature is the *cis* type.

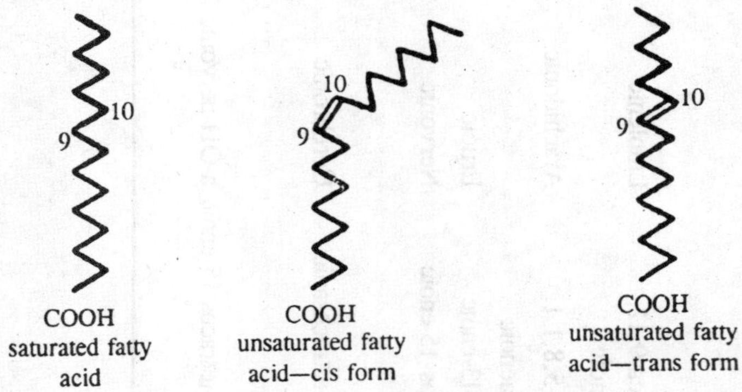

Figure 4.1: Cis and trans forms of unsaturated fatty acid.

As in the case of all aliphatic compounds containing —C=C— groups, the unsaturated fatty acids can undergo addition reactions (e,g., with H_2 or I_2). The double bond is also vulnerable to attack be ozone to yield aldehydic products

$$-CH_2-\overset{H}{C}=\overset{H}{C}-CH_2+O_3 \rightarrow CH_2-\overset{H}{C}\underset{O-O}{\overset{O}{\frown}}\overset{H}{C}-CH_2 \xrightarrow[H_2]{HOH}$$

$$-CH_2-\overset{H}{C}=O+O=\overset{H}{C}-CH_2-+2H_2O$$

Essential Fatty Acids

As discussed in detail elsewhere, the only unsaturated fatty acids that can be made de novo in humans are the monoenic acids, palmitoleic acid and oleic acid. *Linoleic*, *linolenic*, and *arachidonic acids* must be supplied in the diet and are, there-fore, designated as *essential fatty acids*. Although linoleic can be converted to linolenic and arachidonic acids to some extent, all three acids are most generally designated as essential.

Prostaglandins

As described in detail in the *prostaglandins* are a class of fatty acids which are derivatives of prostanoic acid:

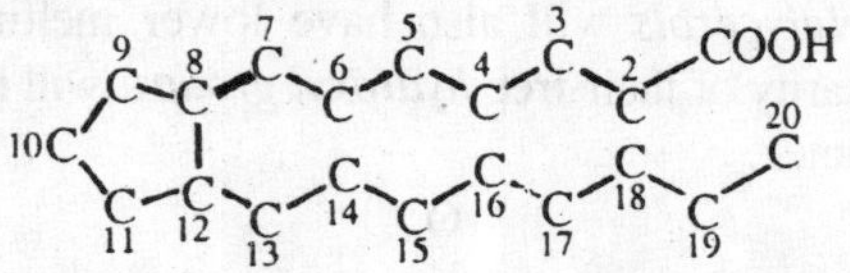

The origin of the prostaglandin family is the essential fatty acid, linoleic acid

Linoleic acid → linolenic acid →

$$\text{arachidonic acid} \xrightarrow[-H_2+O_2]{\text{cyclisation}} \text{prostaglandins}$$

FATTTY ACID ESTERS OF GLYCEROL

Glycerol, a trihydroxyalcohol, is a prominent component of animal lipids because it can form esters with inorganic acids (e.g., *phosphate*) as well as with organic acids such as the fatty acids. The tri-fatty acyl glycerols, the *triglycerides*, constitute the most abundant group of lipids in human tissue. The di-and *monoglycerides* are present in much smaller amounts.

Mono-, Di-, and Triacylglycerols

Three types acylglycerols or glycerides. The acylglycerols can exist in many *isomeric* forms, depending on the number of esterified acyl groups, the identity of the acyl moieties, and their position(s) on the glycerol. For example, a *triglyceride* containing palmitic (P) and oleic (O) acids can have six forms:

PPO, OOP, POP, OPO, OPP, and POO

Carbon 2 of the glycerol becomes asymmetric when the acyl groups on C-1 and C-3this group will be are different. The stereoisomers of this group will be:

C—P	C—O	C—O	C—P
C—P	C—P	C—O	C—O
C—O	C—P	C—P	C—O

The *triacylglycerols* occur as complex mixtures in animal tissues and include fats in which there is only one kind of fatty acid esterified with the glycerol as well as those which contain two or three different fatty acyl groups. In naturally occuring *mixed triacylglycerols*, a *saturated fatty acid* is usually esteri-tiied at C-1 and an unsaturated acid at C-2. The third position may contain either.

The physical characteristics of the triacylglycerols generally reflect the properties of their fatty acid components. Thus, *tripalmityl glycerol* is solid at room temperature, whereas *trioleyl glycerol* is liquid. The mono- and *diacylglycerols* will also have lower melting points and, because of the polarity of their free *hydroxyl groups*, will have significant hydrophilic character.

```
          O
 H        ||
 HC—O—C—R1
 |
 HC—OH            1-or α-Monoacylglycerol
 |
 HC—OH
 H

          O
 H        ||
 HC—O—C—R1
 |
 |        O       1,2- or α, β-Diacylglycerol
 |        ||
 HC—O—C—R2
 |
 HC—OH
 H

          O
 H        ||
 HC—O—C—R1
 |
 |        O
 |        ||
 HC—O—C—R2        Triacylglycerol
 |
 |        O
 |        ||
 HC—O—C—R3
 H
```

Figure 4.2: Mono-, di-, and triglycerides.

Alkyl Ether Acylglycerols

Human neoplasms have been found to contain glycerol derivatives in which the l-, or α-, hydroxyl group is joined in ether linkage to a long *alkyl* or *alkenyl group*. Two examples of such glycerol ethers, are those derived from the alcohols corresponding to stearic and oleic acids. The remaining two hydroxyl groups of the *glycerol* may be esterified with fatty acids. Another class of alkyl ether *acylglycerols*, the *plasmalogens* are found is muscle -oral and *erythrocytes* membranes.

A representative structure of this type of lipid. In addition to phosphate, the polar end of this *amphipathic lipid* usually contains the ethanolamine moiety. The other distinguishing feature of the *plasmalogens* is that one of the two long hydrocarbon tails is contributed by an aliphatic chain in cis α, β-unsaturated ether linkage at position 1 of the glycerol.

Glycosylacylglycerols

Neural tissue has been found to contain a type of l, 2-diacyl-glycerol in which the third hydroxyl group is in glycosidic linkage, usually with galactose as depicted.

Phosphoglycerides

As pointed out in the section on the acylglycerols the pre-sence of just one free hydroxy group in these compounds makes them appreciably polar. The phosphoglycerides constitute a class of *diacylglycerols* whose polarity is even more pronounced because they contain a negatively charged phosphate group. The parent compound of this family of lipids is L-phosphatidic acid, whose structure. In most cases the naturally occurring *phosphatidic acids* contain a saturated fatty acid in position 1 and an unsaturated acyl group on C-2.

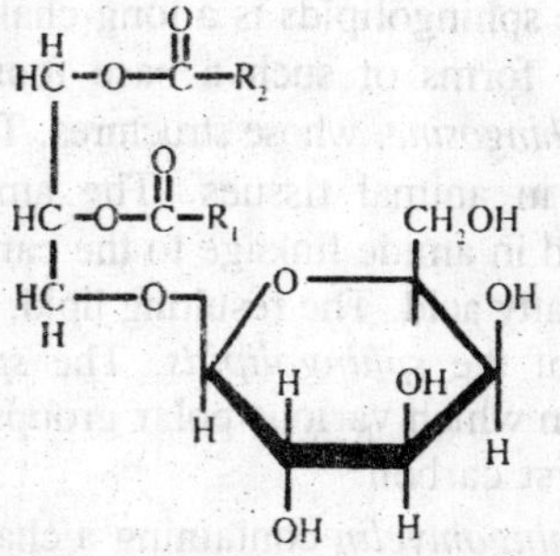

Figure 4.3: A galactosyldiacylglycerol.

The phosphatidic acids occur in only negligible quantities in animal tissues. The most abundant forms of these lipids are those in which the phosphate on C-3 is in a second ester linkage with another alcohol. Such diesters are designated as phosphoglycerides. Particulary prominent among the *phosphoglycerides* of animal tissues are those containing *ethano-lamine*, *choline*, *serine*, *inositol*, *glycerol*, and *glycerol derivatives*. The structures of these compounds. Because of the presence of polar and lipophilic groups within the same molecule, a *phosphatidyl glyceride* will have *amphipathic* character. It is reasonable that these compounds will be abundant in various types of cell membranes. Their detergent properties are also important, as illustrated by the surfactants required for normal lung function. Additional details of structure and function of the phosphoglycerides are presented in the section on lipid synthesis.

Partial hydrolysis products of the phosphoglycerides in which an acyl group is removed from position 1 or 2 of the glycerol are not found in high concentrations in animal tissues. These compounds, known as *lysophosphoglycerides*, injure cells by damaging membranes. The lysophosphoglycerides are the products of specific *phospholipases* which selectively remove acyl groups in *phosphoglycerides*.

FATTY ACID ESTERS NOT CONTAINING GLYCEROL

In this survey of lipid structure which began with a consi-deration of the fatty acids and triacylglycerols, it may be noted that we have progressed from those molecules that serve as forms of metabolic fuel to those that have more specialised functions and that are essential for normal structure in a cell or tissue.

In the more complex lipids described below, the ycerol backbone is replaced by a much longer aliphatic chain. The biosynthesis of these compounds is described else where.

Sphingolipids

The backbone of the sphingolipids is a long-chain, aliphatic amino alcohol. The two major forms of such a base found in humans are sphingosine and *dihydrosphingosine*, whose structures. The C_{18} spbingosines are the most abundant in animal tissues. The amino group of the sphingosine can be joined in amide linkage to the carboxyl of a C_{18}-C_{26} saturated or unsaturated fatty acid. The resulting lipid, called a *ceramide*, is the parent structure of the *sphingolipids*. The *sphingomyelins* are derivatives of ceramide in which various polar groups have been linked to the hydroxyl at its first carbon.

An example of a *sphingomyelin* containing a charged group is one in which either *phosphorylethanolamine* or *phosphorylcholine* is esterified at the ceramide hydroxyl. Like the *phosphoglycerides*, these compounds

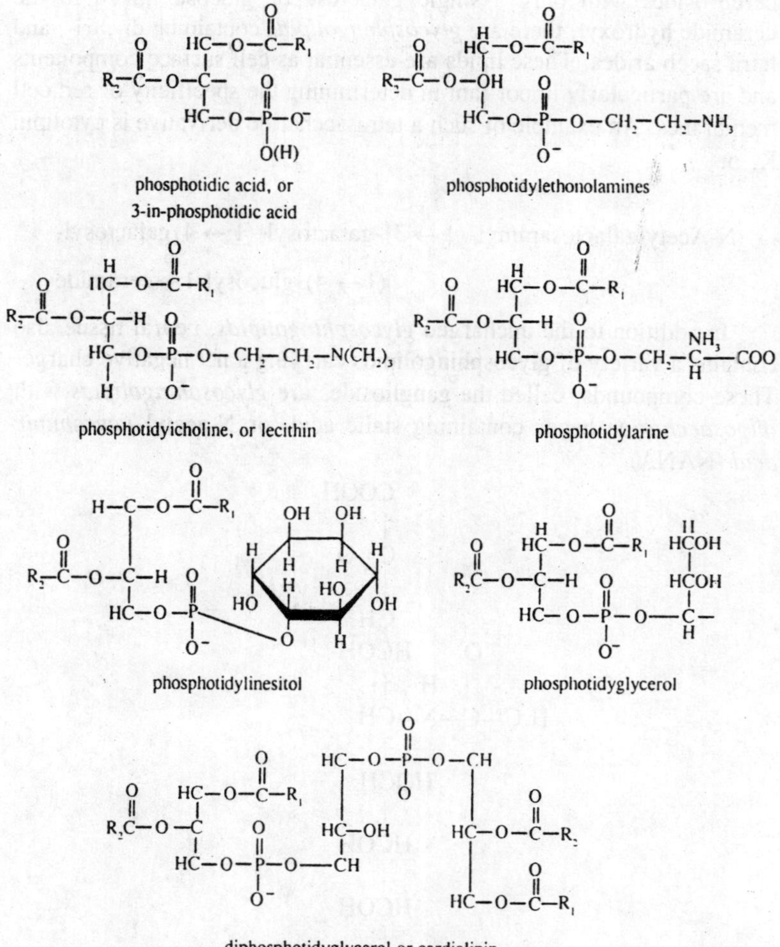

phosphotidic acid, or
3-in-phosphotidic acid

phosphotidylethonolamines

phosphotidyicholine, or lecithin

phosphotidylarine

phosphotidylinesitol

phosphotidyglycerol

diphosphotidyglycerol or cardialipin

Figure 4.4: Structures of phosphatidic acid and representative phospho-glycerides.

are amphipathic. Sphingomyelins containing a hexose in β-glycosidic linkage with the ceramide hydroxyl group are called *cerebrosides* or *neutral glycosphingolipids*.

These complex lipids are most abundant in the myelin sheath of nerves and the *predominant hexose* attached to the *ceramide* is d-*gelactose*. The structure of a typical galactocerebroside is presented in the figure given below. The cerebroside found in nonneural tissues contain d-glucose and are called *glucocerobrosides*. *A* prominant fatty acid found in the *glycosphingolipids* is cerebronic acid. In addition to cerebrosides with only a single galactose or glucose linked to the ceramide hydroxyl, there are *glycosphingolipids* containing di-, tri-, and tetra saccb arides. These lipids are essential as cell surface components and are particularly impor-tant in determining the specificity of red cell membranes. An example of such a tetrasaccharide derivative is cytolipin K, or

$$\text{N-Acetylgalactosaminyl-}(1 \xrightarrow{\beta} 3)\text{-galactosyl-}(1 \xrightarrow{\alpha} 4)\text{galactosyl-}(1 \xrightarrow{\beta} 4)\text{-glucosyl-}1 \xrightarrow{\beta} \text{ceramide}$$

In addition to the uncharged *glycosphingolipids*, neural tissue also contains a variety of glycosphingolipids carrying a net negative charge. These compounds, called the gangliosides are *glycosphingolipids* with *oligosaccharide* heads containing sialic acid, or N-acetyl *neuraminic acid* (NANA)

```
                    COOH
                     |
                    C=O
                     |
                    CH2
       O           HCOH
       ||    H       |
  H3C—C—N—CH
                     |
                   HOCH
                     |
                   HCOH
                     |
                   HCOH
                     |
                   CH2OH
```

Gangliosides can be distinguished by differences in relative amounts and structural relations of d-glucose, *d*-galactose, NANA, and N-acetyl-*d*-galactosamine.

Waxes

Waxes are esters formed between long-chain aliphatic alcohols and fatty acids. An example is a *cetyl palmitata*:

$$CH_3(CH_2)_{14}—C—O—CH_2(CH_2)_{14}CH_3$$

For example, ear wax, or cerumen, contains such esters.

COMPLEX LIPIDS CONTAINING AMINO ACIDS, PEPTIDES AND OLIGOSACCHARIDES

We have been that lipids have the capacity of becoming integrated into complex biological structures. Ultimately, it becomes irrelevant whether the building blocks of such struc-tures are labelled as lipoproteins or proteolipids, as *lipopoly-saccharides* or glycolipids. As illustrations of such hybrid molecules, we shall examine briefly the structures of a *lipo-amino acid*, the *lipopolysaccharides*, and the *lipoproteins*.

Lipoamino Acids

The lipoamino acids and the lipopolysaccharides to be discussed in the next section are particularly abundant in cell membranes and walls of bacteria and are necessary for both protection of the cell and its antigenic specificity. Atypical *lipoamino acid* derivative is that containing lysine. In this structure, the lysine carboxyl is esterified to the position 3' of the glycerol head group of the phosphoglyceride.

Lipopolysaccharides

The rigid basal wall of all bacteria is comprised mostly of a glycopeptide, or peptidoglyean. In gram-negative bacteria this basal layer is covered by a lipopolysaccharide covalently to the polysaccharide and can be removed by mild acid hydrolysis.

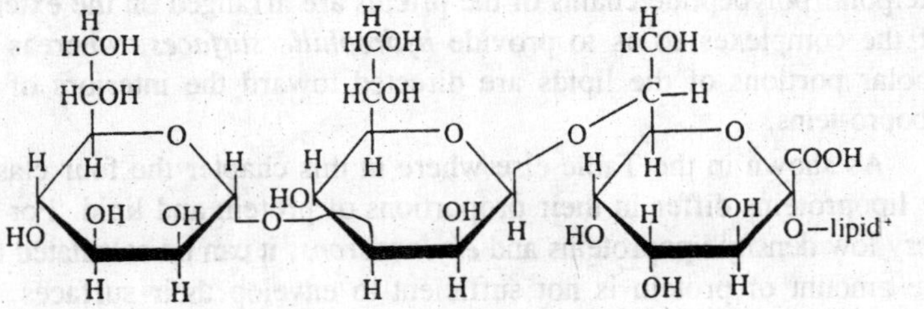

Contains glucosomine, phosphate, acetyl groups and β-hydro--xymyristic acid

$$(CH_3(CH_2)_{10}\overset{OH}{\underset{H}{C}}-CH_2-COOH)$$

Figure 4.5 : Portion of "core" region of a lipopolysaccharide.

The lipopolysaccharides serve as the characteristic somatic O antigens, one of the three kinds of specific surface antigens of coliform bacteria. In their overall structure, these complex lipopolysaccharides consist of a core with "*side-chain whiskers*". Irrespective of differences in their O-antigen specificity, the core regions of the lipopolysaccharides from different strains of a particular bacterium appear to contain a common set of four sugars; *heptose*, *glucose*, *glucosamine*, and *galactose*. The lipid portion is located in the core of the polysaccharide and in *Enterobacteriaceae* has been determined to have the general structure shown in the figure elsewhere in this chapter.

The difference in O-antigen specificity of the bacterial strains can be ascribed to variations in the sugars located in the terminal portions of the side-chain whiskers of the complex. For example, in *Enterobacteriaceae* the outermost regions of the side chains consist of repeating units such as the following:

```
{      abequose                            }
{         |                                }
{  — mannose — rhamnose — glactose         }
```

Lipoproteins

Irrespective of their origin, whether it be the diet or endo-genous sources in the body tissues, lipids are transported in the plasma in association with proteins. Four major classes of lipoproteins of human plasma can be distinguished by *ultracentrifugal* and *electrophoretic* analyses as summarised in the Table elsewhere in this chapter.

The interaction between the lipids and their proteins (apolipoproteins) is noncovalent. The solubility of the resultant *lipoproteins* suggests that the polar polypeptide chains of the prteins are arranged on the exterior of the complexes so as to provide *hydrophilic surfaces*, whereas the apolar portions of the lipids are directed toward the interiors of the lipopro-teins.

As shown in the Table else where in this chapter the four classes of lipoproteins differ in their proportions of protein and lipid. For the very low density lipoproteins and *ehylomicrons*, it can be calculated that the amount of protein is not sufficient to envelop their surfaces. To account for the water solubility of these lipopro-teins, it must be assumed that their *hydrophilic*, polar heads project from the surface of the particles. There are five major groups of apolipoproteins (designated as A, B, C, D, and E), which are distinguished by the lengths of their peptide chains and amino acid compositions.

As shown in Table elsewhere in this chapter the four classes of lipoproteins may share several apolipoprotein components. It should also be pointed out that the four groups of lipoproteins are not homogeneous, one reason being that the A and C families of apolipoproteins consist of two and thrre different polypeptide chains, respectively.

The changes in plasma concentrations of the four classes of

Table 4.2. Composition and physical properties of plasma lipoproteins.

Characteristic	*Chylomicrons*	*Very low density lipoproteins*	*Low density lipoproteins*	*High density lipoproteins*
Density ($g\ ml^{-1}$)	0.92-0.96	0.95-1.006	1.019-1.063	1.63-1.21
Range or average of molecular weights	5×10^8	$5\text{-}100 \times 10^6$	$2\text{-}4 \times 10^6$	$2\text{-}3 \times 10^6$
Diameter (nm)	75-1000	30-75	17-26	7-10
Composition (% of dry weight)				
Protein	1-2	9-10	25	45-55
Triacylglycerols	80-95	55-65	10	3
Phospholipids	3-8	15-20	22-24	21-30
Cholesterol				
Free	1-3	10	8	3-5
Esters	2.4	5	35-37	15
Type of apoprotein				
Major	B, C	B, C, E	B	A, D
Minor	A	A		B, C, E

lipoproteins which reflect the body's need for mobilisation and transport of lipid are described in another section.

NONSAPONIFIABLE LIPIDS

The nonsaponifiable lipids do not contain fatty acids and neither do they yield any other products under the usual conditions of alkaline hydrolysis. The structurally related families of compounds, the terpenes and the steroids, are major nonsaponifiable lipids in animal tissues. The carbon skeletons of both classes are derived from the five-carbon unit, isoprene, or c-methyl-l, 3-butadiene

```
       CH3
   H   |   H
  HC=C———C=CH2
```

Terpenes

Terpene structures may be linear, cyclic, or a combination of both. Presents some representative compounds which illustrate the *hierarchy* of complexity of *terpene structure*. Note that *isoprene* units may be linked head to tail,

```
  C              C
  |              |
C—C—C—C  ↔  C—C—C—C, or tail to tail
                 C              C
                 |              |
            C—C—C—C  ↔  C—C—C—C.
```

Geraniol and *limonene* are typical of the *monoterpenes* found in the essential oils of plants and trees (e.g., geranium and lemon). *Phytol*, containing four isoprene units and referred to as a diterpene, is associated with *chlorophyll*, the *photosynthetic pigment*. Compounds containing six terpene units are designated as *triterpenes*. Squalene, the precursor of cholesterol is an example of this class of terpenes. It will be noted that in squalene the head-to-tail arrangement of the isoprene units is reversed at the center of the molecule, making the two halves mirror images. The higher terpenes include the tetraterpenes, those containing eight isoprene units. Representative of these compounds are the *carotenoids*: β-carotene, shown in the figure is the precursor of vitamin A. Two other fat-soluble vitamins, E and K, are also terpene derivatives.

Another example of a terpenoid derivative essential for human metabolism is ubiquinone, or coenzyme Q. The quinone ring of this electron carrier in the mitochondria is attached to a long isoprenoid side chain.

Steroids

The details of biosynthesis and function of this large group of nonsaponifiable lipids are presented. In this section we highlight some of the structural characteristics of representative steroids which are important in human physiology.

All steroids are derivatives of a fused, tetracyclic ring system, perhydrocyclopentan [α] phenanthrene, whose structure is shown in the Figure. There are six centers of asymmetry in this molecule and substituted derivatives contain more. The ring junctions are usually all trans. Steroids most often contain substituents on C-3, C-11, C-17, C-18, and C-19. In the biosynthesis of the steroids, the first intermediate resulting from the cyclisation of squalene is lanosterol.

Lanosterol is the precursor of cholesterol, which is the most abundant steroid in human tissues. *Lanosterol* and *cholesterol* are commonly referred to as sterols because of the alcoholic hydroxyl at position 3 in figure. The methyl groups at C-18 and C-19 and the -OH at C-3 are all on the same side of the fused ring system and can be regarded as projecting forward toward the viewer. This configuration of the three substituents is referred to as the β orientation.

In some sterols the 3—OH group is oriented α. In structural formulas of the type shown in figure dotted lines usually indicate α-oriented substituents and solid lines are used for β-oriented groups. As discussed in detail elsewhere, the steroids can be grouped according to their physiologic activities. Figure shows the structures of representatives of seven classes of steroids: ***bile acids***, ***glucocorticoids***, ***mineral corticoids***, ***androgens***, ***estrogens***, ***progesterone***, and ***hydroxylated cholecalciferols***.

5

BUILDING BLOCKS

Proteins are the most characteristic chemical compounds found in the living cell. They have high molecular weights and each protein is composed of approximately 20 different kinds of amino acids linked to each other in large numbers. Many proteins contain all of the 20 amino acids and therefore differ from each other only with respect to the number and sequence of their constituent amino acids and with respect to their *threedimensional* configuration.

As a result, the chemical and physical properties of different proteins may differ only slightly and add to the difficulties encountered in separating them from each other and isolating them in the pure state. The amino acids have a great variety of chemically reactive groups, which results in a wide range of reactivity of a protein when exposed to inorganic and organic compounds.

Since such reactions often lead to changes in chemical structure as well in biological activity, chemical contamination of solutions used during the procedure of protein purification must be *rigorously prevented*. A lack of appreciation of the sensitivity of proteins to such chemical reagents often leads to the isolation of a protein in a form different from that present in the cell.

In addition to covalent bonds, which bind amino acids to each other, proteins possess weaker but very important bonds that hold the macromolecule in a unique configuration. Such bonds are quite sensitive to environmental conditions, e.g., excessive stirring of a protein solution in air, exposure to *ultraviolet light*, elevated temperatures, marked changes in pH, and organic solvents.

These procedures lead to alteration of protein structure characterised by loss of solubility and of any biological activity, even though covalent bonds may not have been broken. The protein is said to be *denatured* and frequently the change is irreversible; the *native state* has been destroyed.

Occasionally changes in environmental conditions lead to dissociations of a protein into molecules of smaller size, or of association into larger aggregates Chemical as well as biological properties of the protein are affected by such changes. Despite the difficulties encountered in protein purification,, many proteins have been isolated in the crystalline state with a high degree of purity and biological activity.

A variety of techniques have been developed that take into account the sensitivity of proteins to the usual laboratory manipulations. Possession by a protein molecule of some measurable biological activity, enzymatic or hormonal, simplifies the problem of purification, since we may use an increase in its *specific activity* as a measure of purification. Specific activity may be defined as the activity (enzymatic or hormonal) per unit weight of solids, or, since all enzymes are proteins, their specific activity is usually defined per unit weight of total protein.

Specific activity of the protein enzyme or hormone will increase until no further purification occurs, either because the protein has reached maximum purity or because the procedure has not effected further purification. Since minor alterations in protein structure often lead to loss of activity, isolation of a protein with its activity intact is some assurance that the isolated molecule resembles that which is present in the cell in the native state.

CLASSIFICATION OF PROTEINS

According to Structure

Fibrous Proteins

These proteins are associated with cellular elements and often serve the function of supporting specific structures of the cell. Formerly they were called albuminoids or selerins. Examples are wool, silk fibroin, collagen (connective tissue), myosin (muscle), keratin (hair), and fibrin (blood clot). These proteins are largely insoluble in aqueous media and have high molecular weights that cannot be accurately estimated because of difficulties encountered in their purification.

They appear as fibers made up of linear molecules that are arranged roughly parallel to the fiber axis. They are amorphous and some are capable of stretching and contracting. Human fibrin has the molecular dimensions of 38 × 700 Å.

Globular Proteins

In contrast to fibrous proteins, the globular proteins are soluble in aqueous media and can be isolated in the crystalline state. Although not necessarily *spherical*, they are less asymmetric than fibrous proteins and consist of polypeptides (chains of amino acid residues) that are held together by cross linked groups. Such units are interwoven in a unique fashion and held together in a three dimensional structure by relatively weak bonds.

According to Solubility

Albumins. These are water-soluble proteins that may be precipitated from solution at high salt concentrations, a process called "*salting out*" Decrease in solubility in the presence of high concentration of salts is due to competition between the salt and protein molecules for water, as well as to a decrease in charge on the protein molecules. Examples are egg *albumin*, *serum albumin*, *lactalbumin* (milk), and *leucosin* (wheat).

Globumins

Unlike albumins, these proteins are generally insoluble in salt-free water, and soluble in dilute salt solutions but insoluble in salt solutions at 30 to 50 per cent saturation. With adequate care (low temperature and minimal stirring) the precipitation of globulins or albumins by salting-out procedures need not be accompanied by denaturation. Examples are *serum globulins*, *ovoglobulin* (egg), *myosinogen* (muscle), *edestin* (hemp seed), *amandin* (almonds), and *excelsin* (Brazil nuts).

Glutelins

These are insoluble in neutral aqueous solutions but soluble in dilute acid or alkali. Examples are glutenin (wheat) and oryzenin (rice).

Gliadins (Prolamins)

These are soluble in 70 to 80 per cent ethanol and insoluble in water or absolute ethanol. Examples are *gliadin* (wheat), *hordein* (barley), and *zein* (corn).

Histones

More basic than most proteins, the histones tend to form complexes with acidic compounds in the cell (nucleic acids). They are soluble in water and insoluble in dilute ammonia solution. Examples are globin (hemoglobin), thymus histone (also called nucleohistone since it is found combined with nucleic acids), and scombrone (mackerel).

Protainines

Compared with most proteins, these are relatively small molecules.

The protamines are soluble in water and basic in character. They are found associated with nucleic acids in the sperm of fish, and these are often called nucleoprotamines. Eamples are *salmine* (salmon), *sturine* (sturgeon), *clupeine* (herring), and *cyprinine* (carp).

According to Non-protein Moiety (Conjugated Proteins)

Nucleoproteins

In addition to nucleohistones and nucleoprotamines, there is a third group of proteins capable of binding nucleic acids. These *nucleoproteins* are not basic and hold the nucleic acid by secondary valence bonds. They occur in microorganisms and are soluble in *isotonic salt* solution.

Glycoproteins

These contain carbohydrate moities, including *hexoses*, *hexosamines*, and *hexuronic acids*. Examples are *mucin* (salvia), *osscomucoid* (bone), and *tendomucoid* (tendon).

Phosphoproteins

Examples are *casein* (milk) and *vitellin* (egg yolk).

Chromoproteins

These proteins are pigmented owing to the *prosthetic* non-protein group. They include *hemoglobin* (containing heme, an iron- protoporphyrin), *cerulaplasmin* (copper), *hemocyanin* (copper), *ascorbic acid oxidase* (copper), and *ferritin* (iron).

Dehydrogenases

These are conjugated protein enzymes that possess oxidising activity because of the presence of *prosthetic groups* such as NAD (*nicotinamide adenine dinucleotide*), NADP (*nicotinamide adenine dinucleotide phosphate*), FMN (*flavin mononucleotide*), and FAD (*flavin adenine dinucleotide*).

According to State of Degradation

Native Protein

Proteins in a state presumed to be identical with that present inside the living cell and possessing a unique structure required for biological activity are called native proteins.

Denatured Protein

A protein whose unique native state has been disorganised, so that the original three-dimensional configuration has been converted to that of a random coil is said to be denatured. No chemical groups have been lost from the original molecule.

Derived Proteins

These are products resulting from alterations in the original protein that' involve the removal of amino. acid residues or larger fragments (polypeptides). In some cases, this change does not lead to denaturation or loss of biological activity.

Indeed, the conversion of pepsinogen to pepsin involves the change from an inactive protein to the active enzyme. Hydrolytic cleavage of proteins leads to ill defined products, listed in order of increasing state of degradation primary protein derivatives, *metaproteins*, *proteoses*, *peptones* and *peptides*. These terms have little chemical significance.

PRIMARY STRUCTURE OF PROTEINS

Proteins are made of one or more polypeptide chains. Some proteins contain other kinds of molecules conjugated with the polypeptides, such as carbohydrates or fats, but a substance isn't a protein unless it contains a polypeptide.

The fundamental principle of protein chemistry is that one protein differs from another because each of its polypeptide chains consists of amino acids polymerised in a particular sequence. The amino acid sequences of the constituent poly-peptides are the primary structure of the protein. It is this defined arrangement of amino acid side chains that ultimately gives a protein its properties. Any change in primary structure creates a different protein.

Proteins that differ in only a single amino acid residue at some locations may have nearly identical properties, but there are other locations at which substitution of only one amino for another can cause changes in the nature of the protein that are drastic enough to be lifethreatening if the protein has a critical function.

Number of Different Polypeptides

Although there are only 20 different kinds of amino acids from which proteins are made, there is no danger of exhausting the possible primary structures for the creation of new proteins. A protein made by combining only 100 amino acids molecules is quite small as proteins go, and yet 20 different amino acids can be combined 100 at a time in 20^{100} different ways.

This number is so large that every single protein molecule conceivable in our universe, existing now or at any time in the past, could be unique without using more than an infinitesimal portion of the possibilities. In fact, however, each protein molecule is not unique. Informed guesses at

Table 5.1. Amino Acid Composition of Some Representative Proteins.

Amino acid	*Human Serum Albumin*		*Horse Hemoglobin*		*β-Lactoglobulin*		*Salmine*	
	A	**B**	**A**	**B**	**A**	**B**	**A**	**B**
1	**2**	**3**	**4**	**5**	**6**	**7**	**8**	**9**
Alanine	—	—	7.40	54	7.07	30	1.12	1
Glycine	1.60	15	5,60	48	1.50	7	2.95	3
Valine	7.70	45	9.10	50	5.67	18	3.14	2
Leucine	11.00	58	15.40	75	15.45	44	0	0
Isoleucine	1.70	9	0	0	5.88	17	1.64	1
Proline	5.10	31	3.90	22	5.27	17	5.80	4
Phenylalanine	7.80	33	7.70	30	3.86	9	0	0
Tyrosine	4.70	18	3.03	11	3,69	8	0	0
Trytophan	0.20	1	1.70	5	1.92	3	0	0
Serine	3.34	22	5.80	35	4.07	14	9.1	7
Threonine	4.60	27	4.36	24	5.15	16	0	0
Cystine/2	5.60	32	0.45	2.5	2.29	7	0	0
Cysteine	0.70	4	0.56	3	1.10	3	0	0
Methionine	1.30	6	1.0	4.5	3.21	8	0	0

(Table Contd.)

(Table Contd.)

Amino acid	*Human Serum Albumin*		*Horse Hemoglobin*		*β-Lactoglobulin*		*Salmine*	
	A	**B**	**A**	**B**	**A**	**B**	**A**	**B**
1	**2**	**3**	**4**	**5**	**6**	**7**	**8**	**9**
Arginine	6.20	25	3.65	14	2.88	6	85.2	40
Histidine	3.50	16	8.71	36	1.60	4	0	0
Lysine	12.30	58	8.51	38	11.30	29	0	0
Aspartic acid	8.95	46	10.60	51	11.46	32	0	0
Glutamic acid	17.0	80	8.50	38	19.10	48	0	A0
Amide N	0.88	44	0.87	36	1.07	28	0	0

Insulin (OX)		**Ribonuclease**		**Pepsin**		**Keratin (Wool)**		**Gelatin**	
A	**B**	**A**	**B**	**A**	**B**	**A**	**B**	**A**	**B**
10	**11**	**12**	**13**	**14**	**15**	**16**	**17**	**18**	**19**
4.5	6	—	—	—	—	4.14	46.4	9.3	104.6
4 3	7	1.3	3	6.4	29	6,53	870	26.9	359.0
7.75	8	7.3	9	7.1	21	4.64	39.7	3 3	28.2
132	12	0	0	10.4	27	11.3	863	3 4	26.2
2.77	3	3.1	4	108	28	—	—	1.8	13.8

(Table Contd.)

(Table Contd.)

Insulin (OX)		Ribonuclease		Pepsin		Keratin (Wool)		Gelatin	
A	B	A	B	A	B	A	B	A	B
10	11	12	13	14	15	16	17	18	19
2.9	2-3	3.6	5	5.0	15	9.5	82.6	14.8	129.0
8.14	6	3.6	3	6.4	13	3.65	22.1	2.55	15.5
125	8	7.93	7	8.5	16	4.65	25.7	1.0	5.5
0	0	0	0	2.4	4	1.8	8.8	0	0
5.23	6	12.0	17	12.2	40	10.01	95.4	3.18	30.0
2.08	2	9.0	11	9.6	28	6.42	53.9	2.2	18.5
12.5	12	6.51	8	1.64	4	11.9	98.9	0	0
0	0	0.6	0.7	0.5	2	—	—	0	0
0	0	4.43	5	1.7	4	0.7	4.7	0.9	6.1
3.07	2	5.16	5	1.0	2	10.4	59.7	8.55	49.2
5.21	4	4.22	4	0.9	2	1.1	6.83	0.73	4.71
2.51	2	10.4	11	0.9	2	2.76	18.9	4.60	31.5
6.80	6	14.2	16	16.0	41	7.2	54.1	6.7	50.6
18.60	15	13.0	13	11.9	28	14.1	96.0	11.2	76.1
1.79	12	2.05	22	1.32	32	1.17	83.2	0.07	5.3

the number of different kinds of proteins that a newborn infant makes range between 10,000 and 100,000. (These figures neglect differences caused by partial hydrolysis or other modifications of a polypeptide chain after it is formed).

Assuming there are 40,000 the infant will have on the average about 1×10^{17} identical molecules of each kind of protein. In addition, every one of the $\sim 1.5 \times 10^{8}$ infants born each year will contain identical copies of many of these kinds of protein molecules. Their common humanity is a result of this fact, and it is a marvelous thing that something like 20,000 atoms, to use a typical number, can repeatedly be assembled with so few errors in exactly the same way in each and every human conceived. However, some protein molecules will be different in two is taken at random (but not in identical twins). That is, are many proteins of which one infant will have 10^{17} molecules and the other infant will have none.

The second infant is likely to have molecules of a similar, but not identical, protein in its place. The individuality of the two infants comes from these differences in protein composition. Similarly, the fly that buzzes around the infant will contain many copies of protein molecules that are identical to molecules found in another fly and some that differ in detail.

However, it is almost certain that the human and the fly have no protein molecules in common; their functions are too different to be served in exactly the same way. Proteins are made to satify these diverse needs by varying the way in which the 20 kinds of amino acids are put together, that is, by altering the primary structure, and specification of the variations is a major part of the heredity of the organism.

AMINO ACID COMPOSITION OF PROTEINS

In order to determine the quantity of each amino acid in a protein, the latter is subjected to hydrolysis, using acid, alkali, or enzymes, depending on the amino acid to be estimated. *Acid hydrolysis*, in the absence of air, is usually perferred.

Alkali hydrolyses the amide groups of *asparagine* and *glutamine*, destroys *cystine* and *arginine*, and leads to racemisation. Enzymes are not only slow in their action out the hydrolysis may be incomplete, and the presence of the enzyme protein may interfere with the analysis. Some of the analytical methods that have been devised follow.

Chemical Methods

The earliest, and from the quantitative point of view, the **crudest** method involved the direct isolation of amino acids or their derivatives, based on solubility differences. The introduction of specific colour reactions for many amino acids replaced the less quantitative isolation procedures but did not provide highly specific methods for all amino acids.

Microbiological Methods

Many microorganisms can be grown in media that are fairly simple and whose composition can be varied at will. These microorganisms often require particular amino acids (essential amino acids) for growth. Within certain limitations a direct relationship can be demonstrated between the quantity of such an amino acid in the medium and the ability of the organism to grow (divide). Sometimes, evidence of a specific metabolic activity—e.g., lactic acid production is measured.

A *standard* growth curve (or acid production curve) can be constructed to show this relationship for specific quantities of the amino acid. By comparison of this standard curve with the growth of the organism in a medium that lacks the amino acid under examination but to which has been added varying quantities of a neutralised protein hydrolysate, we can calculate the quantity of amino acid in the hydrolysate.

This method has been quite helpful but suffers from several limitations. Amino acids that are available as standards are not always pure, microorganisms often substitute one amino acid for another, and the reaction of an organism to specific quantities of an amino acid often differs in the presence of other amino acids.

Isotope Dilution Method

This method can be extremely accurate, but it requires pure isotopically-labeled amino acids, it is time-consuming, and sometimes as is the case with stable isotopes, it requires costly assay equipment (mass spectrometer).

In principle, an amino acid containing an isotope of known specific activity is dissolved in a known quantity of a protein hydrolysate, and the same amino acid is then isolated and purified, either as such or as a derivative. It is not necessary to isolate the amino acid quantitatively, but rather to be certain of its purity.

If we assume that the tagged amino acid that was added has the same solubility as the nonisotopic amino acid in the hydrolysate, a comparison of the specific isotope activities of the amino acid that was added and the one that was isolated will reveal the extent of dilution of the added acid by that present in the bydrolysate. For example, the amount of glycine originally present in a hydrolysate (y) can be calculated from the amount of glycine added (x), and its content of N^{15} $(C_x)_y$ as well as the N^{15} content of the isolated glycine (C):

$$y = \left(\frac{C_o}{C} - 1\right)x$$

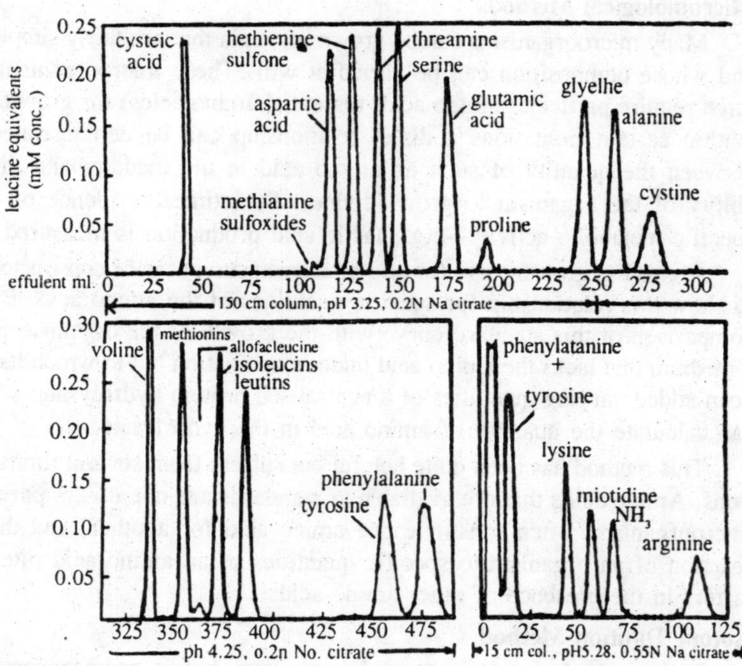

Figure 5.1: Chromatographic fractionation of a synthetic mixture of amino acids on columns of Amberlite IR-120.

Chromatographic Analysis

By far the most useful and grecise method for determining the quantity of each amino acid in a protein hydrolysate is the separation of the amino acids on a column of adsorbent, followed by the consecutive elution of each acid. As little as 2 mg of pure protein is required for a complete analysis.

The amino acids are eluted separately by buffer solutions of varying composition. The solutions, as they leave the column, are separated into small fractions and ninhydrin reagent is added to each for colourimetric estimation of amino acid content. By the use of known amino acids and standardisation of conditions, the position of each amino acid in the effluent can be determined. Figure elsewhere in this chapter illustrates the use of this technique. Automatic devices are now incorporated into an apparatus that simplifies the determination. The quantity of each amino acid is calculated from the area under each curve.

SEQUENCE DETERMINATION

A knowledge of the number of amino acid residues of each kind per molecule of protein does not reveal the sequence in which these amino acid residues appear in the polypeptide chain. Techniques for the determination of amino acid sequences will now be discussed.

N-Terminal Amino Acids

A single polypeptide chain contains one N-terminal amino acid residue whose amino group is usually free. Sanger devised a technique for identifying this residue based on the reaction of the protein with 2, 4-dinitrofluorobenzene in mildly alkaline solution at room temperature. The reaction proceeds as follows:

R
|
CH.CO—NH.etc + [F, NO_2, NO_2 benzene ring] → CH.CO—NH. etc. (R above; NH below, bonded to a ring bearing two NO_2 groups) + HF
|
NH_2

N-Terminal end of a polypeptide — DNFB — DNF-polypeptide

The N-terminal residue now contains the dinitrophenyl (DNP) tag. The DNP-protein is hydrolysed, liberating the N-terminal DNP-amino acid and other amino acids in the free state (except for lysine, whose E-amino group would be free to react with the reagent). If lysine were the N-terminal residue, di-DNP lysine would be liberated. The yellow DNP-amino acid or acids are separated chromatographically and identified. If more than one polypeptide chain is present, more than one different N-terminal DNP amino acid may be element.

Another method for identifying the N-terminal residue, and one that is more helpful for the analysis of small peptides because amino and residues can be removed from the Nterminal position one at a time. The peptide is reacted with phenylisothiocyanate and then treated with acid, liberating the N-terminal residue as the phenylthiohydantoin, together with the peptide, which now has a new N-terminal residue. The reaction is illustrated as follows:

$$\begin{array}{l} R.CH.NH_2 \\ \quad | \\ \quad C{=}O \\ \quad | \\ \quad NH \\ \quad | \\ \quad etc. \end{array} + C_6H_5.N{=}C{=}S \longrightarrow \begin{array}{l} \qquad\qquad\quad S \\ \qquad\qquad\quad \| \\ R.CH.NH.C.NH.C_6H_5 \\ \quad | \\ \quad C{=}O \\ \quad | \\ \quad NH \\ \quad | \\ \quad etc. \end{array} \xrightarrow{H_+}$$

$$\begin{array}{l} \quad R.CH\text{——}NH \\ \qquad | \qquad\quad | \\ \quad O{=}C \qquad C{=}S \\ \qquad\quad \diagdown \;\; \diagup \\ \qquad\qquad N \\ \qquad\qquad | \\ \qquad\qquad C_6H_5 \end{array} + \begin{array}{l} NH_2 \\ | \\ etc. \end{array}$$

C-Terminal Amino Acids

Several methods for determining the nature of the C-terminal residue are available. One of these is based on the specificity of the enzyme carboxypeptidase, which preferentially hydrolyses the peptide linkage involving the C-terminal residue. It is necessary to examine the free amino acids liberated at various intervals in order to determine which amino acid is first liberated, because as thus is amino acid residue is removed, another C terminal residue appears.

Another method for identifying the C-terminal residue involves the conversion of the terminal carboxyl group to the primary alcohol by means of $LiAlH_4$ and $NaBH_4$, followed by hydrolysis and isolation of the amino alcohol and its identification.

Polypeptide Cross-linkages

Although the basic unit of some proteins contains but one polypeptide chain, other basic units contain two or more. In either case the polypeptide

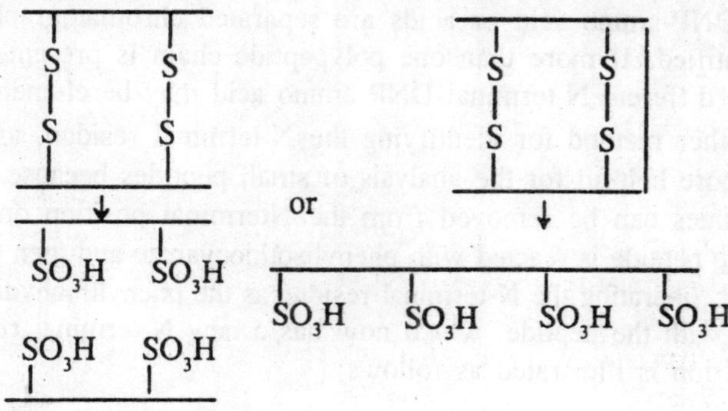

chains may be held together by disulfide groups of two half-cystine residues. Treatment of then protein with performic acid results in oxidation of the disulfide to two sulfonic acid groups that are then part of different cysteic acid residues. The modified polypeptide chains can then be separated. Disulfide bridges can also connect portions of the same polypeptide chain, but their cleavage by performic acid oxidation does not give rise to several polypeptides.

Intermediate Amino Acid Sequences

Completion of the primary structural analysis of a protein requires and elucidation of the sequence of residues between the N-terminal and C-terminal residues of each polypeptide chain. We can illustrate this technique with insulin, whose complete primary structure was the first to be determined (Sanger).

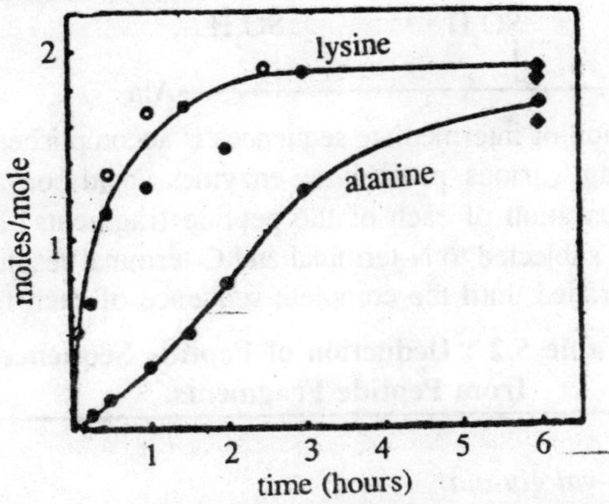

Figure 5.2: The kinetics of amino acid release from rabbit muscle enolase *during digestion with carboxypeptidase.*

The unit monomer of insulin has a molecular weight of approximately 6000. Application of the Sanger method of N-terminal amino acid a ialysis results in the isolation of two N-terminal amino acids- glycine and phenylalanine. The C-terminal amino acids are asparagine and alanine. The insulin molecule contains three disulfide bridges.

Treatment of insulin with performic acid results in the production of two altered polypeptides—A and B—which can be separated. The A chain has glycine and asparagine as N- kind C-terminal residues respectively, whereas the B chain has phenylalanine and alanine as N- and C-terminal residues, respectively. The altered A chain contains for cysteic acid residues, whereas the B chain has two such residues. Reconstruction of the original molecule from these data suggests the preliminary structure elsewhere in this chapter.

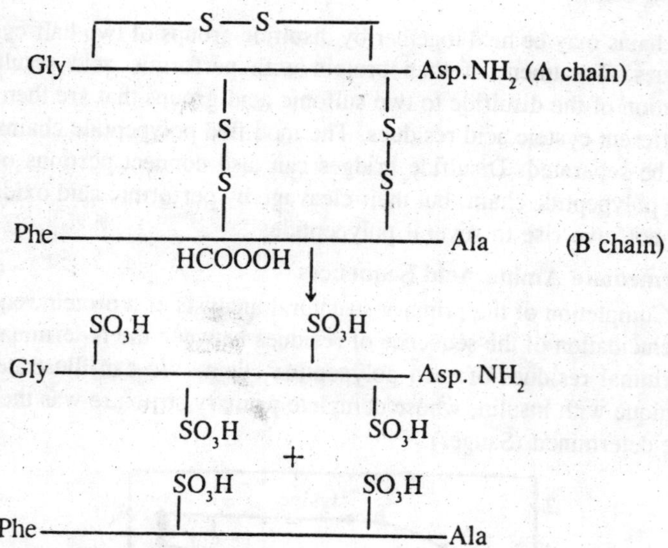

Determination of intermediate sequences is accomplished by partial hydrolysis, using various proteolytic enzymes, acid, or alkali, and followed by separation of each of the peptide fragments. Each small fragment is next subjected to N-terminal and C-terminal residue analyses and further degraded until the complete sequence of each fragment is

Table 5.2 : Deduction of Peptide Sequence from Peptide Fragments.

Dipeptides

ser-his leu-val glu-ala

his-leu val-glu

Tripeptides

ser-his-leu

leu-val-glu

val-glu-ala

ala-ltu-tyr

tyr-leu-val

Higher peptides

ser-his-leu-val-glu

val-glu-ala-leu leu-val-CySO$_3$H—gly—

Deduced sequence

—ser-his-leu-val-glu-ala-leu-tyr-leu-val-CySO$_3$H—gly—

determined. As examination of the sequences in all of the fragments show enough overlapping of residues to enable one to reconstruct the sequences in the original polypeptide.

Figure elsewhere in this chapter shows the complete sequence of residues in the insulin molecule. It consists of two polypeptide chains held together by disulfide bridges.

Figure elsewhere in this chapter shows the complete amino acid sequence for the first enzyme molecule whose primary structure was clarified-ribonuclease. This molecule, which contains 124 residues, consists of one polypeptide chain, with lysine and valine as the N-terminal and C-terminal amino acids respectively. The disulphide bridges of eight half cystine residues connect different portions of the chain resulting in a structure that has several restrictions for the three-dimensional folding of the molecule.

Figure elsewhere in this chapter shows the complete amino acid sequence in -cytochrome c from horse heart. It is of interest for two reasons -the N-terminal amino acid glycine is ace.,lated (also true for tobacco mosaic virus protein, whose N-terminal residue is acetyl serine), and the two cysteine residues are attached to the vinyl side chains of the porphyrin by thioether bonds.

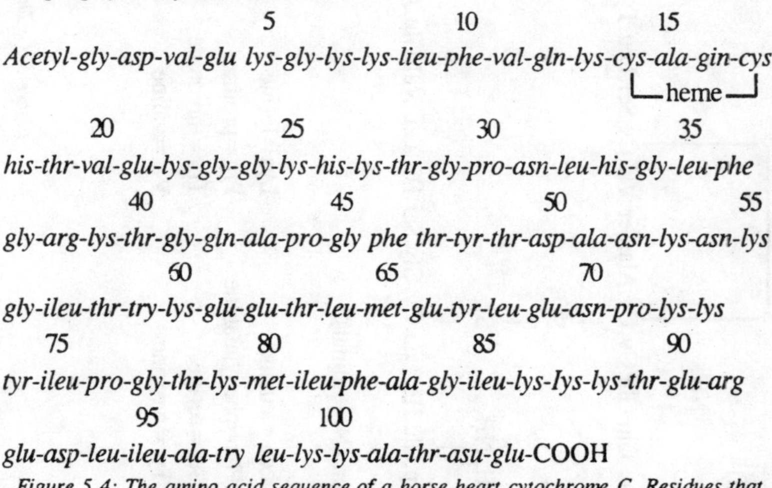

5 10 15
Acetyl-gly-asp-val-glu lys-gly-lys-lys-lieu-phe-val-gln-lys-cys-ala-gin-cys
heme

20 25 30 35
his-thr-val-glu-lys-gly-gly-lys-his-lys-thr-gly-pro-asn-leu-his-gly-leu-phe

40 45 50 55
gly-arg-lys-thr-gly-gln-ala-pro-gly phe thr-tyr-thr-asp-ala-asn-lys-asn-lys

60 65 70
gly-ileu-thr-try-lys-glu-glu-thr-leu-met-glu-tyr-leu-glu-asn-pro-lys-lys

75 80 85 90
tyr-ileu-pro-gly-thr-lys-met-ileu-phe-ala-gly-ileu-lys-Iys-lys-thr-glu-arg

95 100
*glu-asp-leu-ileu-ala-try leu-lys-lys-ala-thr-asu-glu-*COOH

Figure 5.4: The amino acid sequence of a horse heart cytochrome C. Residues that differ in human heart protein are indicated by italics.

Amino Acids Replacements in Hemoglobin

The hemoglobins have served as an excellent source of a pure protein that can be obtained from an almost unlimited number of individuals from various around the world and that. can be analysed for

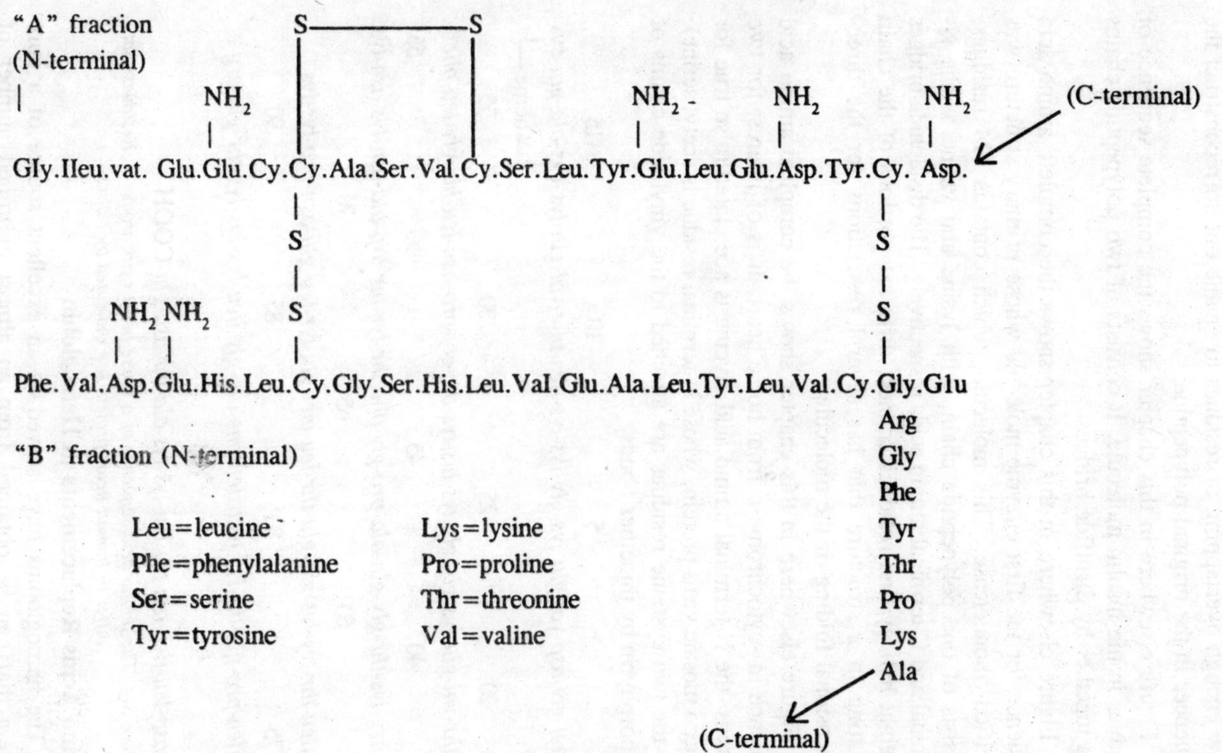

Figure 5.3 : The structure of beef insulin.

differences in amino acid sequences. Normal adult human hemoglobin consists of four polypeptide chains—two α- and β-chains.

It may therefore be designated by the symbols $\alpha_2{}^A\beta_2{}^A$, where A identifies the chains as belonging to adult hemoglobin (HbA). Each α-chain consists of 141 amino acid residuesfwith valine and arginine as the Nterminal and C-terminal residues respectively. Each of the pchains contains 146 residues, with valine and histidine as N-terminal and C-terminal residues. The intermediate sequences for both chains have been identified.

Sickle cell anemia is a disease characterised by the "*sickling*" of red cells induced by lowered oxygen tension. The hemoglobin in such cell appears to aggregate, its solubility is decreased, and the cell is removed from the circulation at an abnormally rapid rate, leading to severe anemia. The disease is hereditary, and severe sickle cell anemia occurs in individuals *homozygous* for an abnormal gene. It also occurs as a sickle cell trait, in the absence of an anemia, in carriers who are heterozygous, having one normal and one abnormal gene.

The demonstration of an abnormal hemoglobin (Hb S) in red cells from individuals with the disease was made by Pauling, Itano, Singer, and Wells. It was shown that the hemoglobin prepared from individuals with this disease possessed an electro-phoretic mobility, when measured at a pH alkaline to its IEP, lower than that of normal adult hemoglobin. Pauling called sickle cell anemia a "*molecular disease*", to replace the earlier designation "inborn error of metabolism" suggested by Garrod for the inherited disease alcaptonuria.

As information about the amino acid sequence in hemoglobin developed, it became apparent that the differences between Hb A and Hb S was slight. The exact differences was narrowed to the replacement of a glutamic acid residue at position 6 of the β-chain in Hb A by a valine residue in Hb S. The reason for the lower mobility of Hb S was clear, since the negatively charged glutamic acid residue had been replaced by an uncharged valine.

Detection of such a minor difference is somewhat easier than might be anticipated, because of the use of a "*mapping*" or "*finger print*" technique. The protein (or polypeptide chain) is treated with a proteolytic enzyme, such as trypsin, which hydrolyses the chain at specific positions. The resulting peptide mixture is separated by the use of paper electrophoesis and chromatography. When trypsin acts on Hb A and Hb S, the results in shown Figure are obtained. The peptide fragments appear as sports produced by reaction with ninhydrin. Close examination of the two "*maps*" reveals that Hb S has one peptide that occupies a new and different position from the corresponding peptide in Hb A.

In a duplicate chromatogram (not treated with ninhydrin), the spot is cut from the paper, and the peptide is eluted and subjected to further analyses to reveal its exact amino acid sequence. Peptide No. 4 is an octapeptide having the following sequences:

Hb A val-his-leu-threo-pro-*glu*-glu-lys

Hb S val-his-leu-threo-pro-*val*-glu-lys

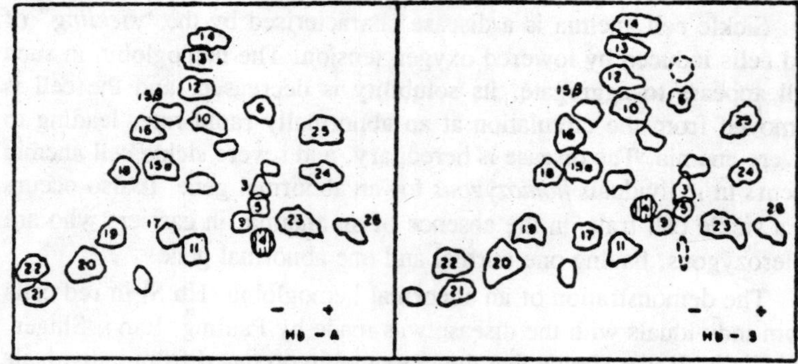

Figure 5.5: Peptide fingerprints of hemoglobins A and S.

This technique has made it possible to examine the hemoglobin of large numbers of individuals, and many instances of "abnormal' hemoglobins have been discovered. These all appear to have a single amino acid replacement, either in the a-chain or the 9-chain. They are summarised in Table elsewhere in this chapter.

Note that Hb C has a lysine residue at position 6 of the 13-chain where Hb A has a glutamic acid residue and Hb S has a valineresidue. Consequently, Hb C has an electrophoretic mobility lower than that of Hb S. Hb C is found in people who are simultaneously heterozygous for S and C hemoglobins. The disorder is much milder than sickle cell anemia. Simultaneous heterozygosity for Hb C and thalassemia (Mediterranean anemia) as also associated with clinical symptoms and abnormal blood cell.

Hemoglobins that contain an amino acid replacement are called 'abnormal because they differ from normal adult Hb A. Compared with the normal population, people with abnormal hemoglobin are rare, and it is quite certain that the alteration in structure represents a gene mutation. However, not all abnormal hemoglobins are associated with clinical manifestations of disease.

Table 5.3 : Amino Acid Replacements in the Hemoglobin of Humans.

	α-Chain						*β-Chain*					
	16	30	57	58	68	116	6	7	26	63	67	125
Hb A	lys	glu	gly	his	asn	his	glu	glu	glu	his	val	glu
Hb I	asp											
Hb												
GHonolulu		gln										
HbNorfolk			asp									
Hb MBoston				tyr								
Hb GPhiladelphia					lys							
Hb Oindonesia						lys						
Hb S							val					
Hb C							lys					
Hb CSan *Jos*								gly				
Hb E									lys			
Hb MSaskatoon										tyr		
Hb MMilwaukee											glu	
DPunjab												gin
HbZurich										arg		
Hb OArabia												lys

Since the oxygen-carrying power of hemoglobin is crucial to life in animals that possess hemoglobin, it is obvious that mutations would be lethal if they resulted in alterations greatly affecting this oxygen-carrying power. Later in this chapter we illustrate how an understanding of the molecular structure of hemoglobin, in primary sequence as well as in three-dimensional conformation, can be used to develop a concept of the function in this molecule.

Amino Acid Sequences in Normal Hemoglobins during Development

Examination of hemoglobin isolated from the blood of the developing fetus has revealed the presence of several polypeptides that differ from the α- and β-chains found in adult hemoglobin. Illustrates the appearance of these polypeptide chains in fetal hemoglobin. Soon after the fetus begins to develop, there is a rapid production of α- and ε-chains. The α-chains persists. whereas the ε-chain disappears, and a new polypeptide, the γ-chain makes its appearance. During the late stage of fetal growth the β-chain increases in quantity as the γ-chain decreases correspondingly. The composition of fetal and adult hemoglobins is illustrated by the following symbols:

Hb A (Adult) $\alpha_2^A\beta_2^A$

Hb F (Fetal) $\alpha_2^A\gamma_2^F$

Hb A_2 $\alpha_2^A\delta_2^A$

Our knowledge about the amino acid sequence in the ε-chain is still incomplete. In all of the fetal henmoglobins, the α-chain remains a constant constituent. The β-, γ-, and δ-chains each contain 146 amino acid residues and closely resemble each other with respect to their amino acid sequences. There are only 10 differences between the β- and δ-chains and approximately 39 differences between the β- and γ-chains. Figure shows the entire sequence of the γ-chain, as well as those amino

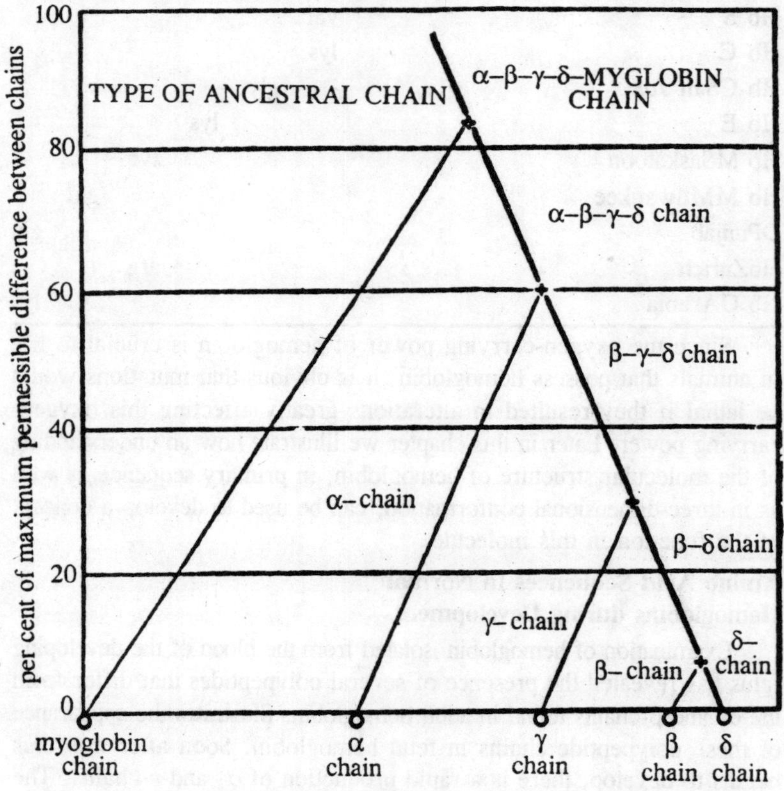

Figure 5.6 : Hemoglobin-myoglobin relationship is traced back through evolution, based on the number of differences in the various chains.

acids residues that are replaced in the β-chain. We see that the three-dimensional configurations of the sand β-chains of normal adult hemoglobin are very similar although the amino acid residues in these two chains differ in number.

If the amino acid sequences of the two chains are compared, it is possible to identify 65 residues that are the same, providing one is permitted to allow spaces between certain residues (deletions) and one assumes that insertions have been made during their development from some common ancestral polypeptide. This is considered likely since all the chains serve the same biological function and possess similar conformations, in addition to the similarities in their amino acid sequences.

In considering the possibility of the evolution of present polypeptide chains of hemoglobin from a common ancestral polypeptide, we must take into account the protein myoglobin of muscle, which also binds molecular oxygen reversibly. This protein has a molecular weight roughly one-quarter the molecular weight of hemoglobin, and it consists of but one polypeptide chain with 153 amino acid residues. Although the sequences in human myoglobin are not yet completey determined, while myoglobin has been shown to possess 37 residues that are the same as those in the α- and β-chains of humanhemoglobin. It is therefore likely that myoglobin has also evolved from the same ancestral polypeptide as the hemoglobin chains. Figure illustrates this hypothesis, which suggests tha mutations and gene duplication were responsible for th changes in amino acid sequences. Figure illustrates a computation by Pauling of the age of the ancestral chains.

Amino Acid Sequences in Similar Molecules from Different Species

Among the interesting consequences of our ability t determine the primary structure of a protein molecule is th opportunity to note any differences between similar molecules from different species.

Insulin serves as a good example since complete amino acid sequences have been determined for a relatively large number of species. Table elsewhere in this chapter illustrates the differences that have been detected. Amino acid residues for certain sequences in bovine insulin are shown for comparison; the numbers refer to the position of the amino acid residue, starting with 1 for the Nterminal amino acid. Note that the B-chain for cod insulin contains one more residue than that of bovine insulin, whereas bonito (2) has one less residue. Its terminal residue for the B-chain in lysine.

The posterior lobe of the pituitary produces two hormones - vasopressin and oxytocin-which are octapeptides. These were isolated in the pure state, their structures were determined and they were then synthesised by du Vigneaud and collaborators. They are shown below

Oxytocin (bovine)

```
┌────────S—S────────┐
│                   │
Cy—tyr—ile—gln—asn—Cy—pro—leu—gly
```

Vasopressin (bovine)

```
┌────────S—S────────┐
│                   │
Cy—tyr—phe—gln—asn—Cy—pro—arg—gly
```

Vasopressin (hog)

```
┌────────S—S────────┐
│                   │
Cy—tyr—phe—gln—asn—Cy—pro—lys—gly
```

Amino acid differences in hemoglobins of several different species are shown in Table elsewhere in this chapter.

Table 5.4 : Amino Acid Differences in Hemoglobins of Different Species.

	56	57	58	59	60	61	*(α-chain)*
Human A	lys	gly	his	gly	lys	lys	
Bovine					ala		
Sheep					glu		
Goat					glu		
Horse		ala					
Orangutan		asp					

Melancyte Stimulating Hormones (MTH)

Colour differences among different species are due to differences in the quantity and distribution of the pigment *melanin*. This compound is derived from the amino acid tyrosine. Melanin is part of the colour in skin and hair, and in the "*ink*" released by the squid. The pituitary glands of many animals have yielded two forms of the polypeptide hormone MSH that promote rapid skin colour changes by affecting the melanocytes-cells that are derived embryonically from nerve tissue. One of these, α-MSH appears to be same in the pig, cow, horse,

1 2 3 4 5 6 7 8 9 10 11 12 13 14
H.Ser.Tyr.Ser.Met.Glu.His Phe.Arg.Try.Gly.Lys.Pro.Val.Gly
HO Phe...............Pro.Try.Val.Lys.Val.Pro Arg.Arg Lys Lys
39 24 23 22 21 20 19 18 17 16 15

Figure 5.7 : Primary sequence of amino acid residues in ACTH.

monkey, and probably man, whereas β-MSH varies somewhat from one species to another. Their amino acid sequences are shown in Figure.

Amino Acid Sequence and Biological Activity

The relation of enzyme activity to the amino acid sequence around the "active site" is discussed under enzymes.

An interesting study of biological activity of a polypeptide hormone and its amino acid sequence has been reported for the adrenocorticotropic hormone (ACTH), a straight-chain polypeptide containing 39 residues. Snythesis of large units of this chain have revealed that : (1) The smallest unit that has biological activity is the N-terminal tridecapeptide, corresponding to residues 1—13. (2) Further elongation of the chain result in increased activity, with maximum potency attained by a chain of 20 residues.

The "*active site*" of the polypeptide lies in the N-terminal tridecapeptide, whereas residues 14 to 20 appear to increase the affinity of the effective subunit for the receptor of the cell.

SECONDARY STRUCTURE OF PROTEINS

The spatial arrangement of polypeptide chains is now known in many proteins through the use of x-ray crystallo graphy. Even casual inspection shows that the chains are not bunched up in random masses; most proteins contain many segments with recognisable patterns. These segments of regular geometry are called secondary structures.

Three types of secondary structures are common in protein: *helixes, reverse turns,* and *pleated sheets*. It is the sequence of amino acid side chains that determines whether a particular portion of *the polymer* will fold into one of these patterns, but there are some general considerations that apply to all.

Structure will be favoured in which there is a minimum of unfilled space. *Dispersion forces,* also called *London forces,* stabilise structures in which atoms are tightly packed. Either the structure itself must be snughly together or it must have crevices large enough to be crammed with solvent molecules.

A structure will be favoured if it brings all of the C=O and N—

H groups of the polypeptide backone into positions where they can form *hydrogen bonds* with each other or with side chain groups. Otherwise, the polypeptide would tend to open up so that these groups could form hydrogen bonds with water. Put another way, the protein tends to fold so that the energy released by internal hydrogen bonding of the back bone groups is as great as the energy that could be released by bonding with water.

BONDS

The forces that are involved in folding a protein molecule are briefly summerised in the following sections.

Covalent Bonds

The interaction between two atoms arising from the sharing of an electron pair produces a strong interatomic force-the covalent bond. Such forces may be involved in holding two nolvueptide chains together or two section of a single polypeptide chain in a folded configuration-

Table 5.5 : Dissociation Energies to Typical Bonds.

Bond	*DissociationEnergy (ev.)*
H—H	4.40
>C—C<	2.55
>C—H	3.80
>C—N<	2.13
>C—N (peptide)	3.03
>C=C<	4.35
>C=O	6.30
—C≡C—	5.35

for example, the crosslinking disulfide bond contributed by two half-cystine residues. Such bonds are not easily broken, but they may be converted to sulfhydryl groups by reductions, resulting in a separation of the two polypepide chains or a unfolding of a single chain.

Ionic Bonds

These are due to the charged groups present in the various amino acid side chains, the extent and the kind of charge varying with pH. These are illustrated in Figure. They interact by ,mutual attraction, giving rise to specific conformations of the protein.

Van der Waals Forces

These forces are due to the mutual attraction of nonpolar hydrocarbon side chains of amino acids, like valine and leucine. These forces attract when the atoms are at a certain distance from each other, but repel when the atoms approach each other more closely.

The forces are determined by the distribution of electrons around the two atoms. By means of x-ray diffraction analysis of a crystal, one can measure the distance separating the atoms and assign a van der Waals radius to the atom. This distance is the closest approach that an atom can make to a neighbouring atom without forming a chemical bond. The energy of such forces is about 1 kcal./mole.

Steric Effects

These effects are caused by the presence in amino acid residues of bulky side chains, which can prevent the close.approximation of groups or chains.

Hydrogen Bonds

These bonds are the result of the affinity of hydrogen atoms for electronegative atoms such as N or O. Electronegative atoms have a tendency to share electrons with H.

$$\overset{-}{\text{—N}}\text{—}\overset{+}{\text{H}}\ldots\ldots\overset{-}{\text{O}}\text{=}\overset{+}{\text{C}}\text{—}$$

In this illustration the attraction of the H atom for the 0 atom results in a *hydrogen bond* between the amide and carbonyl groups that are constituents of the peptide bond in proteins. Weak hydrogen bonds are electrostatic in nature, whereas strong hydrogen bonds are shorter and have a greater covalent character.

The energy of such bonds is less than 10 kcal./mole, but because there are a large number of peptide bonds in a protein molecule, hydrogen bonds assume great significance in stablisation of the three-dimensional structure of the molecule. We may visualise two ways in which such bonds might serve to restrict protein structure: (a) by holding two polypeptide chains together, or (b) by holding sections of.a single polypeptide chain in specific orientation in space.

Hydrogen Bonds

(1) Between peptide groups:

```
      /              /
 R—C              C=O
      \            /
      C=C......H—N
      /              \
 H—N               C—R
      \             /
```

(2) Between neutral groups of side chains:

```
    O.........HO
   /            \
 —C              C—
   \            /
    OH.........O
```

(3) Between charged and neutral groups of side chains

```
                         O
                      ⊖ ||
 —⟨benzene⟩—OH.........C—
                        /
                       O
```

(4) Between charged groups of side chains

```
                  O
               ⊖ ||
 —NH3⊕ ........C—
                 /
                O
```

(5) Between side chains and peptide groups

```
                          \
                           C—R
                          /
 —⟨benzene⟩—OH......O=C
                          \
                           N—H
                          /
```

Hydrophobic Bonds

(1) Between two nonpolar side chains groups.

$$—CH_3.......... H_2C—$$

(2) Between nonpolar side chains and backbone groups

```
       \
  R—C—H......H3C—
       /
  O=C
      \
       N°—H
      /
```

Figure 5.8: Examples of hydrogen bonds and hydrophobic bonds in proteins.

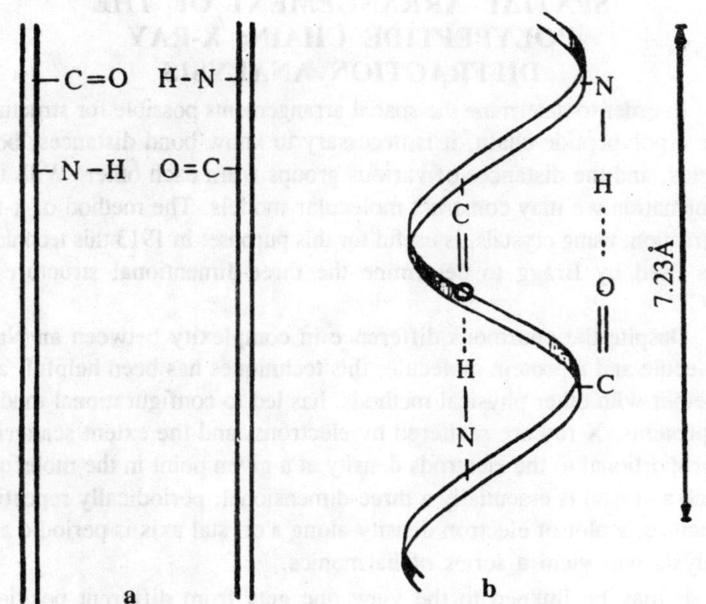

Figure 5.9.

The concept of the hydrogen bond, when applied to protein structure, helps to explain why there is often an alteration of the three-dimentional structure, and in the case of biologically active proteins, a loss of activity, despite the use of procedures that leave covalent bonds intact. An example is the denaturation of proteins. The breaking of a hydrogen bond in proteins involves the absorption of some 10 kcal./mole. Nevertheless, in a medium that has a high dielectric, whose molecules themselves have a marked tendency to form hydrogen bonds, the solvent molecule e.g., water causes disruption of the hydrogen bond with the absorption of only 1.5 kcal./mole.

H H
C=O... H− N+ O... H→C =O...H +N−H...O
H O O H
H H

Effective denaturing agents for proteins are urea and guanidine. The action of urea as a hydrogen bond breaker has been questioned, but it has been supposed that the effectiveness of such compounds of denaturing agents is associated with their ability to disrupt hydrogen bonds, causing an unfolding of the polypeptide chains.

SPATIAL ARRANGEMENT OF THE POLYPEPTIDE CHAIN; X-RAY DIFFRACTION ANALYSIS

In order to determine the spatial arrangements possible for structures like a polypeptide chain, it is necessary to know bond distances, bond angles, and the distances of various groups from each other. With this information we may construct molecular models. The method of x-ray diffraction, using crystals, is useful for this purpose: in 1913 this technique was used by Bragg to determine the three-dimentional structure of NaCl.

Despite the enormous difference in complexity between an NaCl molecule and a protein molecule, this techniques has been helpful, and together with other physical methods, has led to configurational models of proteins. X-ray are scattered by electrons, and the extent scattering is proportional to the electrods density at a given point in the molecule. Since a crystal is essentially a three-dimensional, periodically repeating structure, a plot of electron density along a crystal axis is periodic and analysis will yield a series of harmonics.

It may be linkned to the view one gets from different positions along a roadway when looking into a field of regularly spaced trees in an orchard. The complete diffraction pattern of a crystalline protein consists of a regular threedimensional array of spots, any plane of which can be sampled by means of an x-ray photograph. Each of the spots in the diffraction pattern corresponds to a summation of component waves of the periodically-repeating electron density distribution of that crystal.

Each wave has a characteristic amplitude, wave length, and phase. With relatively simple molecules, the solution to the problem can be accomplished by use of a mathematical technique known as *Fourier series,* which involves an equation with a number of variables with respect to their effects from zero to infinity. A complication in the analyses was an inability to locate some primary reference point in the protein molecule. This was solved by the method of *isomorphous replacement* the insertion of a large heavy ion, such as Hg^{++}, which is identified by its high electron density. Density contour maps can now be plotted for any sampled plane.

By plotting these maps on clear plastic and stacking them, a three-dimensional picture showing regions of high and low density is obtained. Application of this method to proteins requires an enormous number of calculations in order to get the fine resolution required to identify features of the molecule. High-speed computers are required to get a resolution of 2Å, which involves the treatment of 10,000 reflections. The importance of resolving power in an analysis by this method lies in the fact the distance between covalently bonded atoms is 1 to 1.5 Å, whereas atoms of neighbouring groups brought together by forces such as hydrogen bonding or van der Waals forces lie 2.8 to 4 Å apart. A resolution of 2 Å would separate polypeptide chains or groups of atoms, whereas 1 A would allow a separation of individual atoms. Actually, 1.4 Å is just distinguishable by this method; below this the diffraction pattern fades away.

Helixes

Polypeptides, like any long repeating polymer, tend to form coils. Why is this so? Because any repetitious displacement in space travels a spiral path. To eliminate the effects of differing side chains, imagine a polypeptide made from one kind of amino acid. Execpt for those residues near the ends of the chain, each will be exposed to much the same environment as its neighbours.

Therefore, they will tend to form the same Ramachandran angles relative to each other; the polypeptide backbone will regularly berd in the same way. It so happens that there are particular spiral arrangements with sterically favourable Ramachandran angles that are reasonably compact and also enable all possible hydrogen bonds to be formed with the polypeptide backbone.

The α-Helix

The most common helix in protein occurs when the polypeptide chain twists into a right-handed screw with the N—H group of each amino acid residue hydrogen. bonded to the C=O group of the fourth following residue, which is in the adjacent turn of the helix. This screw is the α-helix. It is a rod-like structure coated with amino acid side chains extending outward from its axis. While it is obviously suited for creating long fibre-like molecules, we shall see that it also commonly occurs in globular protein molecules.

Other helixes can be formed by single polypeptide chains so as to hydrogen bond all possible backbone group. All that must be done in principle is to tighten or loosen the coils of the helix to align the bonding groups at other residues in the chain. For example, the chain can be tightly twisted so that each residue forms hydrogen bonds with the third successsive residues, rather than the fourth Such a helix has exactly three residues per turn, rather than the 3.6 found in the cc-helix. It is called a 3_{10} helix because each hydrogen bond creates a 10membered ring and there are three residues per turn. (If the twisting is somewhat less tight, an N—H the C=O groups of both the third and the forth following residues, forming a ture of α_{11}-helix.)

Similarly, untwisting the α-helix so as to align groups on the fifth following residues will create a π-helix.

Of all of these possibilities, the a-helix is the most common. Why? Because it is the one in which the twist is just tight enough to pack the backbone snugly without straining the bonds or leaving a hollow core. Even so, forces occasionally favour localised formation of these other helixes.

Reverse Turns

Polypeptide chains often fold back upon themeselves so as to change, or even reverse, direction. Important forms of these bonds involve hydrogen bonding form one residue to the third following residue. One of six ways in which this can be done is shown in the figure. There are other specific geometries that do not involve hydrogen bonding within the bend, but in either case these "*reverse turns*" or "*chain reversals*" occur in specific favoured configurations.

Pleated Sheets

Compact sheet-like structures form when segments of polypeptide chains lie side by side at nearly maximum extension. These β-structures, like the α-helix, are made with all possible hydrogen bonds between the segments of backbone. They are also known as pleated sheets because of the zigzag appearance they have when viewed on edge.

The sheets may be antiparallel, with the chain running in opposite direction in adjacent segments. This is the form taken when a single chain is successively folded back upon itself in the most compact possible way. The sheets also may be parallel; this requires that the polypeptide chain be looped back upon itself in some way so that adjacent segments of the sheet run in the same direction, creating a β_P structure.

Sheets in proteins are often formed with both parallel and antiparallel segments, and frequently are somewhat twisted. These configurations are represented by a convention in which the general direction of the backbone is shown by a flat arrow, pointing from the amino terminal of the chain toward the carboxyl terminal. The figure also illustrates the resemblance of a commonly occurring pleated sheet, when depicted in this way, to a design motif often found on ancient pottery—the Greek key.

The striking characteristic of the pleated sheet is that side chains

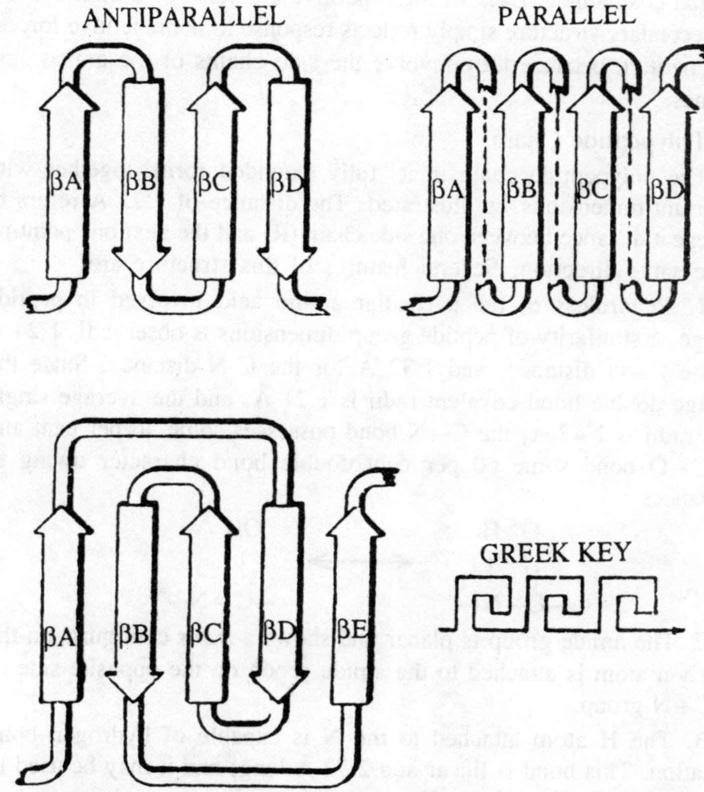

Figure 5.10: The course of the Polypeptide chain in pleated sheets is represented diagrammatically by broad flat arrows pointing toward the carboxyl terminal of the chain.

of successive amino acid residues appear on opposite sides of the sheet; the sheets therefore tend to form when every other side chain can interact in some effective way. The intervening side chains are segregated on the opposite side of the sheet.

A small irregularity known as the β-bulge is sometimes introduced into pleated sheets by insertion of an extra amino acid residue into one segment. The result is a greater distortion of the sheet enabling the formation of some structures to be discussed shortly.

Nonordered Arrangement

Part of a polypeptide chain may lack any recognisable geometrical order of the kinds just described and is sometimes said to be in a random coil. This is a misleading term because the arrangement of a polypeptide chain is seldom random, even if it appears as tangled as a crushed coathanger. Lack of the repetitive elements of design that we call secondary structure simply reflects response to nonrepetitive forces-nonrepetitive because they involve the side chains of the amino acid residues.

The Polypeptide Chain

The polypeptide chain in its fully extended form, together with important dimensions, is illustrated. The distance of 7.23 Å refers to the repeat distance between one side chain (R) and the next one pointing in the same direction. Several features of this structure are:

1. Regardless of the particular amino acid involved in peptide linkage, a similarity of peptide group dimensions is observedL 1.24 Å for the C—O distance, and 1.32 Å for the C-N distance. Since the average double bond covalent radii is 1.21 Å, and the average single bond radii is 1.47 Å, the C—N bond possesses some 40 per cent and the C—O bond some 60 per cent double bond character owing to resonance:

```
 O  H              OH
 ||  |    <----->    |
—C—N—             —C=N—
```

2. The amide group is planar and shows a *trans* configuration-the α-carbon atom is attached to the amide group on the opposite side of the C—N group.

3. The H atom attached to the N is capable of hydrogen bond formation. This bond is linear and 2.72 Å long; and it may be used to interconnect C=O and N—H groups along the same chain or two chains to each other.

Stable Configurations of the Polypeptide Chain

If we consider the close packing of polypeptide chains and the effect of bulky side chains of many of the amino acids, it becomes clear that different kinds of configurations are possible. One way to approach this problem is to examine synthetic polymers of single amino acids like polyglycine or poly-L-alanine. Generally there are two kinds of stable configurations possible. These are characterised by hydrogen bonding between adjacent polypeptides, and hydrogen bonds within the same polypeptide chain.

Hydrogen Bonding between Adjacent Polypeptides

One form of polyglycine has a configuration consistent with that of planar polypeptide units lying in a common plane to form a sheet structure.

Silk fibroin, a protein containing 44 per cent glycine and 40 per cent alanine plus serine, has a structure similar in many respects to polyglycine, but with the polypeptide chain less extended, in order to make room for the side chains of amino acids other than glycine. The chains are arrayed alongside in an antiparallel fashion and bent to give the appearance of a pleated sheet. The polypeptide bankbone has a zigzag rather than a straight form, and the C=O and N—H groups are oriented so that all H bonds can be formed by aligning a series of these polypeptides with alternate chains moving in opposite directions. The successive carbonyl and amid groups are hydrogen bonded to chains on either side of the original chain.

Although most of the properties of silk fibroin can be understood from this simplified structure, this model does not account for all the amino acids found in the protein, including many with extensive side chains, that would act to disturb the model and produce minor complications seen in the experimental diffraction pattern.

Keratin is a fibrous protein that differs from silk fibroin in that it contains nearly all the known amino acids. It has a high cystine content and is low in histidine, methionine, and tryptophan. Unlike silk fibers, wool and hair are easily stretched. Natural keratins in the unstretched form are designated *α-keratins,* whereas the stretched variety are the *β-keratins*. Data obtained from x-ray diffraction studies, together with one of the polypeptide configurations suggested by Pauling and Corey, lead to a model of *β-keratin* best described as a parallel chain pleated sheet, similar in some respects to silk fibroin, but differing because of the more bulky side chains. Although the problem of the structure of *β-keratin* is far from solved, the working model is close to the actual structure.

Hydrogen Bonds within the Same Polypeptide Chain

α-keratin, sections of globular proteins, and some forms of synthetic polypeptides are folded or coiled in a more compact configuration than can be explained by semiextended polypeptides. To explain such coiling, Pauling and Corey, on the basis of known interatomic distances and bond angles and the study of models, suggested the *α*-helix configuration. Important features of this structure are

1. Each NH group is connected to a C=O group by a hydrogen bond at a distance equivalent to three amino acid residues.
2. Each amino acid residue is 1.47 to 1.53 Å distant from the next residue, and the helix makes a complete turn for each 3.6 to 3.7 residues. The x-helix is only one of many different helices

possible and is determined by the number of residues per turn. For example, the x-helix would have 4.4 amino acid residues per turn.

3. The vertical distance along the axis from any point on the helix to a corresponding point directly above it is approximately 5.4 Å (3.6 × 1.5) and is referred to as the pitch of the helix.
4. With 3.6 residues per turn, five complete turns correspond to 18 residues, yielding a repeating interval of approximately 27 Å (1.5 × 18), which has been found in diffraction patterns made of fibers of polyamino acids of certain kinds. There are two ways of coiling a helix—a right handed and leftbanded screw. The right-handed helix has been shown to be more stable.

In the case of a-keratin, agreement of experimental x-ray diffraction reflections with those that have been calculated is less clear than those obtained from synthetic polyamino acids, although the 1.5 Å reflection is strong. Instead of a spacing. of 5.4 Å, corresponding to the pitch of the α-helix just described, a spacing of 5.1 α is obtained for α-keratin and in a slightly different position. These differences can be reconciled if we assume that the α-helices of α-keratin are coiled into bundles or cables of "*coiled coils*". Similar structures may be used to picture the configuration of other members of the keratin group-myosin of muscle, epidermin of skin, and fibrinogen of plasma.

The α-helix is a closely packed structure with no open spaces in the center. Inside this structure is the repeating part of the polypeptide chain, exposing all amino acid side chains to the surface. When proline residues are incorporated into the right-handed α-helix, little disturbance in Structure need occur.

However, if these residues are inserted into the lefthanded α-helix, the sequence is disrupted, and they may serve as a turning point for the helix-it will reverse direction. In collagen-the principal fibrous protein of mammalian connective tissue, tendon and bones—there is one proline or hydroxyproline per four residues, and this protein represents a special problem in configuration. It is suggested that collagen is constructed of left-handed helical structures for each of three polypeptide chains. These three left-handed structures twist around each other forming a large right-handed helix.

TERTIARY STRUCUTURE

At this level of organisation of the molecule, a protein has size and shape as well as a *unique arrangement* of its polypeptide chains. To gather such structural information requires the application of a variety of physicochemical methods that can be used with protein solutions, as well as the x-ray diffraction method, which can only be applied to the crystal. A brief account of these methods is given.

Two Examples

The tertiary structure of a polypeptide chain is the complete form assumed by the chain—the combination of various secondary structures and nonordered segments into an ordered whole. We shall concentrate for now on globular proteins and consider the more specialised fibrous proteins as we come to them. Outlines the polypeptide backbone in each of two proteins. The one on top is part of liver alcohol dehydrogenase, a protein that is responsible for the oxidation of ethanol to acetaldehyde. The one on the bottom is the polypeptide of myoglobin (muscle hemoglobin), a protein responsible for binding and transporting oxygen in muscle fibers.

Although both of these proteins are relatively globular, their overall shapes give no clue to the quite different arrangements of the polypeptide chains creating the proteins. The liver alcohol dehydrogenase contains a central core of a bent parallel pleated sheet made from six segments of chain enclosed by four recognisable segments of α-helix. (The segments are lettered according to the order in which they appear from the amino terminal of the chain. αA is the first segment used to form a pleated sheet, (αA is the first segment appearing in an α-helix, and so on). Between these segments are various twists and bends.

The myoglobin polypeptide, on the other hand, contains no pleated sheet and is made up almost entirely of segments of α-helix, with what appears to be a minimal amount of connecting chain between some segments.

Patterns of Tertiary Structure

The two examples of tertiary structure we used are not at all unusual. The general arrangement shown for liver alcohol dehydrogenase is similar to one found in several other proteins, some of quite different function. The same arrangement of stacked helixes seen in myoglobin is also present in other oxygen-binding proteins, ranging from a hemoglobin found in the nitrogen-fixing nodules of legumes to circulating human hemoglobins.

Indeed, is now possible to recognise several ways in which groups of pleated sheet and helix segments are organised to make globular proteins. Some polypeptide chains fold into stacks of pleated sheets surrounded by bends and nonordered segments. In others, a large pleated sheet is folded into a cylinder (the β-barrel). Many others have pleated sheet cores surrounded by helixes, much like alcohol dehydrogenase, but this can be accomplished in many ways. (The term *super secondary structure* has been applied to these recurring patterns, but they do not have the fixed geometry of the fundamental secondary structures.)

Domains

One polypeptide chain need not be folded into one compact unit. For example, in alcohol dehydrogenase part of the chain is folded to make the structure. However, another part of the chain is folded into a similar but not identical structure with a separate function even though it is covalently liked to the first structure by the connecting region of the polypeptide chain. Domains, then, are recognisable units of tertiary structure resembling separate proteins but formed from one polypeptide chain. They often fold independently of each other and retain their character if the chain linking them is broken.

ORIGIN OF THE TERTIARY STRUCTURE

Why do liver alcohol dehydrogenase and myoglobin differ so much in the conformation of lowest energy content? We know the difference in amino acid composition of the two proteins is the root cause, exactly how do the forces developed by the varying side chains generate the structure? This is a question we cannot answer in detail: it is not yet possible to deduce exactly how a polypeptide chain of known primary structure will fold, but great progress has been made toward the development of the general principles.

Secondary Structures are Preferred

Since the pleated sheets and helixes saturate the backbone with hydrogen bonds, permitting it to be buried away from water, it is not surprising that many polypeptides preferentially fold into a combination of these shapes, often connected by specific kinds of reverse turns. Even those globular proteins without recognisable sheets or helixes, and they are not common, are mainly made from specific turns.

The Structures are Compact

The same forces that cause secondary structures to be compact also cause them to combine into snug tertiary structures. Openings are usually more than one atom wide and arc intended for access of solvent or for combination with other molecules The protein may have bizarre forms, but it has little unoccupied space. It is not porous.

The Surface is Polar and the Interior is Nonpolar

One of the major forces affecting the shape of a protein comes from the tendency of polar side chains to associate with water and of nonpolar side chains to agglomerate in the interior of the molecule. These effects stabilise contortions of the polypeptide that bring the hydrophilic residues to the surface and bury the hydrophobic residues.

Hydrophilic Side Chains

The most hydrophilic amino acid residues are those with net charges, such as aspartate and glutamate with their negatively charged carboxylate groups, and lysine and arginine with their positively charged ammonium and guanidinium groups. The chain must nearly always fold so as to keep them in contact with water.

It is only possible to bury these groups when the charge in neutralised by forming the uncharged forms, combining positive and negative groups, or by *dissipating* the charge through multiple hydrogen bonds. Even when this is done, the folding must be strongly favoured elsewhere to remove these polar groups from contact with water. The carboxamide groups of glutamine and asparagine are also polar enough to strongly favour conformations that keep them in contact with water.

Hydrophobic Bonds

The force causing nonpolar side chains to huddle together out of contact with water comes not from any specific attraction of the groups for each other but from the propensity of surrounding water to resist being forced into ordered structures by the presence of the nonpolar groups. The so-called hydrophobic bond is really an association created by spurned water, and it is the same kind of force that causes insoluble oils to form compact spheres upon suspension in water. How does it arise?

Molecules of water have a strong tendency to form hydrogen bonds with each other as is shown by the release of heat when water freezes into the fixed lattice of ice. There is an opposing tendency of all molecules to stay dispersed, to be random and unordered. This tendency, measured as entropy, is increased by a rise in temperature. The two effects just balance at the melting point. At higher temperature, hydrogen bonds persist for shorter and shorter times; each molecule becomes more and more promiscuous.

A nonpolar molecule introduced into liquid water represents a eunuch at an orgy. Since it cannot form hydrogen bonds or other bonds created by polarity, neighbouring water molecules lose some of their possibilities for random union, and associate with each other longer than they would in the absence of the nonpolar intruder.

However, this longer association represents greater order, a step toward freezing into ice, and we have already said that creation of such order at temperatures above the melting point can be accomplished only by the introduction of energy. It follows that dispersing nonpolar molecules in water requires energy; if they are left to themselves, they diminish their exposed surfaces by clustering, thus impeding fewer random liaisons among water molecules. They become a less effective moral minority, as it were.

Some Nonpolar Chains are on the Surface

If the degree of polarity of the side chains were the only determinant of structure, polypeptide would fold so that every group not interacting with water was buried in the interior. The tendency to make this separation is very strong and can aid in shaping complex surfaces, but the separation is not complete.

There are usually more nonpolar side chains than can be accommodated in the interior of a protein, and some are exposed on the surface of most proteins. This is an important part of the development of specific function, as well as shape. Having more reactive groups separated by the relatively unreactive nonpolar side chains gives individual character to the surface geometry—the stars are bright because the sky is black.

GENERATION OF SPECIFIC STRUCTURES

The *geacrat nature* of the forces acting on a polypeptide chain seem clear, but it is more *difficult* to understand exactly what influence a particular amino acid sequence may have on protein conformation. However, we have some clear-cut examples with which to make a start.

Proline is a Helix-Breaker

This is the one apparently infallible general rule for prelication of structure. A residue of proline cannot occur in the piddle of a straight α-helix. The nitrogen atom of a prolyl group has no attached hydrogen atom with which to bond a preceding carbonyl group, and the fixed orientation of the bulky side chain gets in the way of any preceding turn. Therefore, a prolyl group can only be in the first turn of an α-helix; the preceding amino acid residue must occur in some other configuration, at least bending, and usually breaking, the helix.

```
       H2
       C
      / \
   H2C   CH2
      \ /
       N—C...H
         \
          C...
          ‖
          O
```

prolyl group

Reverse Turns are often Exposed

What causes chain reversals? The chain can be bent upon itself at any residue if the bend enables strong bonding elsewhere, but there are some amino acids whose side chains favour reverse turns. Proline, with its odd geometry, is an example. Glycine is another; its lack of any side chain permits some tight folds that are not possible with other amino acids, and it is more likely to occur in a reverse turn than any other:

```
   H  H
   |  |
...N—C—C...
      |  ‖
      H  O
```

glycyl group

Other amino acids favour the formation of reverse turns because they have hydrogen-bonding groups only a short distance removed from the peptide backbone. These include asparagine, serine, and aspartate

```
                                                    H
                                                   /
                          ...O  O...          ...O  N
                              \ /                 \ / \
      O—H...                   C                   C   H...
     /                         |                   |
   H  CH2                   H  CH2              H  CH2
   |  |                     |  |                |  |
...N—C—C...               —N—C—C—            ...N—C—C..
   |  ‖                     |  ‖                |  ‖
   H  O                     H  O                H  O
```

seryl group aspartyl group asparaginyl group

Here we have it: reverse turns are likely to contain polar side chains or to have a nearly naked backbone where glycine occurs. Either event will favour exposure to water at the surface of the molecule. The elbows of protein molecules are often bared.

Pleated Sheet or α-Helix?

Since the pleated sheets and helixes have complete bydrogen bonding of the backbones, the forces favouring one over the other must be developed by the differing dispositions of the side chains. For example, generation of a secondary structure with a hydrophobic surface that can be faced toward the interior may facilitate construction of a compact molecule. We can sometimes predict if formation of a pleated sheet or α-helix is more likely to occur by picturing what happens in the two cases.

Alternate amino acid side chains are exposed on the same face of a pleated sheet. Therefore, if we plot the amino acid sequence as a zigzag, those residues to one side will be on one face of a sheet, those on the other side on the other face. The left part of Figure shows what happens when this is done with the sequences of the two longest segments of pleated sheet from liver alcohol dehydrogenase together with the preceding and following residues. Folding these sequences as a sheet creates a massive aggregation of hydrophobic groups from the side chains of valine and phenylalanine:

H_3C CH_3 / H CH / ...N—C—C... / H O (=O)

valyl group

C_6H_5 / H CH_2 / ...N—C—C... / O (=O)

phenylalanyl group

The other face of the sheet has smaller or charged side chains. Also notice the nature of the residues at the two ends of the sequences. The bends at these positions are loaded with glycyl, prolyl, aspartyl and asparaginyl groups. Here we have unusually clear examples of segments of primary structure that will make sharply defined hydrophobic regions through formation of segments of pleated sheet, terminated on each end by an abundance of chainreversing residue.

Generation of an α-halix

Since an a-helix has 3.6. residues per turn, the residues are spaced at 100° intervals viewed from the end, as one goes around the helix. We therefore can get an idea of the forces that will be generated by drawing "*helical wheel*" in which -the positions of successive residues in a chain are set down at 100° intervals around a circle. We are in effect looking down the cylindrical axis of the helix and observing the distribution of side chains around the surface. Figure gives a comparison of plotting the same primary structure as a pleated sheet and as a helical wheel. The thing to look for is some indication of the development of specific bonding forces along particular sides of the helix. Could, for example, one side of the helix be facing the hydrophobic interior and another side be facing the aqueous phase? If so, a helix becomes more likely.

When we look at the segments in those terms, they appear rather nondescript. The segments from 12 o'clock are relatively polar; the remaining portion is nonpolar, but the groups are too widely scattered to be definitive. The first segment couldn't be a continuous α-helix in any case because of the prolyl residues that end it. The second segment possibly might be, but the apparent bonding forces are nothing like those seen with a pleated sheet configuration.

Now let us look at a sequence from myoglobin. Here the pleated sheet configuration results in a rather random distribution of groups, while the wheel shows that an a-helix will create a well-defined region of bulky hydrophobic groups. In fact, this segment of the chain is a portion of an α-helix, with the hydrophobic segment in contact with other parts of the chain in the interior, and the remainder of the surface being exposed to surrounding water. It is the potential for creating this distribution of groups that makes myoglobin fold in the way it does.

THE HEME POCKET

There is another important aspect to the structure of myoglobin. It is a heme protein, meaning that it belongs to a class of conjugated proteins that contain heme groups in addition to polypeptide chains, and the primary structure must be arranged so as to hold the heme in position.

Heme is combination of iron (II) and protoporphyrin 1X. We shall say great deal more about the porphyrins later. The point of importance now is that the porphyrins contain a highly resonant planar ring system in which four nitrogen, atoms are fixed in the centre at a spacing that is ideal for bonding of their unshared electrons with a metal ion. The porphyrins therefore act a tetradentate (four-toothed) ligands, and protoporphyrin IX has an especial affinity for iron.

Iron (II) ions have a coordination number of six, which means that they can associate with six electron pairs atom. When a ferrous ion forms a chelate with a porphyrin, two of its coordinations positions are still unfilled, so free heme in, aqueous solution will be hydrated as shown in the figure. One of the functions of the polypeptide portion of myoglobin. (the "*globin*") is to provide a histidine residue at a position where one of its nitrogen atoms will link to the fifth coordination position of the ion. The sixth position of the iron is left open and now has the ability to bind oxygen.

The remainder of the peptide chain wraps around the porphyrin ring to hold it in position. The walls of this heme packet are created by hydrophobic side chains front several of the surrounding helical segments, closely fitting the hydrophobic prophyrin ring. In addition, the aromatic side chain of a phenylalanyl group lies paralled to the porpbryin for interaction of the π-electrons. The porphyrin has two charged side chains, but these protrude out of the herne pocket to contact the aqueous phase.

In sum, the polypeptide chain of myoglobin has an amino acid sequence that favours the formation of a particular set of a-helices that will snugly nestle around a heme group and hold it in four ways: chelation of the iron, hydrophobic exclusion of water, interaction of aromatic rings, and close packing of the atoms. Indeed, the polypeptide is built to fold into a particular tertiary structure in the presence of heme. (Otherwise, the hydrophobic heme pocket would be exposed to water).

SEDIMENTATION IN THE CENTRIFUGAL FIELD

Development of the ultracentrifuge by Svedberg in 1925 marked a milestone in modern biochemistry. Its subsequent modification has led to its general use in the biochemistry laboratory. Since the rate of movement of a protein molecule is a function, among other things, of its size, one can determine the apparent molecular weight/of a protein in solution under a variety of conditions. Two methods used are sedimentation velocity and sedimentation equilibrium.

Sedimentation Velocity

By means of a suitable optical system it is possible to measure the rate of movement of protein molecules in a centrifugal field of high intensity in a solvent (e.g., aqueous buffer) less dense than that of the protein solution. If a single molecular species is present, a single bondary between the moving protein molecules and the buffer will be observed. If several species of protein molecules are present, these will be observed as a series of boundaries, each boundary moving at a different rate, depending on its molecular weight.

Calculation of the molecular weight involves a knowledge of the sedimentation constant s, the diffusion contant D, density of the medium d_m and the partial specific volume V R and T are the gas constant and absolute temperature respectively. This relationship follows:

$$M = \frac{RT}{D(1 - Vd_m)} . (s)$$

where

$$s = \frac{dx/dt}{x\omega^2}$$

and *dx/dt* is the rate at which the particle travels at distance x from the center of rotation, and w is the velocity of revolutions in radians per second (angular velocity). The diffusion constant is obtained by observing the rate of motion of the protein boundary in the absence of a centrifugal field.

The partial specific volume is determined by observing the density of solutions of the protein as a function of concentration of the anhydrous protein. The sedimentation constant s is characteristic of the molecule under stated conditions and varies from 1×10^{-13} second to 200×10^{-13} second for proteins. Because of the negative exponents, another units, the Svedberg (S), is used. This is a unit of sedimentation constant equaling 1×10^{-13}, so that proteins may be said to have values of 1 S to 200 S.

Sedimentation Equilibrium

In this method, we observe the distribution of protein molecules in cell in which the gravitational field is less than that used for the sedimentation velocity method. The molecules distribute themselves at equilibrium according to size, and average molecular weights can be calculated over the range force 1000 to 10,000.

Light Scattering

The scattering of light is proportional to the weight-average molecular weight, and the proportionality factor can be experimentally determined. The sensitivity of this method increases with the molecular weight. In addition, this method can yield information regarding the shape of molecules. Molecules of dimensions of 200 to 300 Å show interference because of the light scattered from different parts of the same molecule. As a consequence, the scattering is diminished, the effect increasing with the scattering angle. For example, cellulose nitrate has a rather broad distribution of molecular weights, and the size is indicative of an extended random coil. On the other hand, tobaccomosaic virus shows rods of 300 A in length for a molecular weight of 39.5 million, with an effective diameter of 150 A.

OSMOTIC PRESSURE

The pressure required to prevent the diffusion of solvent through a membrane into a solution of a protein dissolved in that solvent is termed osmotic pressure. The osmotic pressure of a protein solution is a property that gives a macroscopic response with dilute macromulecular solutions, measuring the number of molecules of solute per unit volume of solution. (vowing the weight of material per unit volume of solution, ve can determine the number of molecules and thus get a lumber-average molecular weight.

Table elsewhere in this chapter lists the molecular weights of several proteins determined by various methods.

OPTICAL ROTATION

As was noted earlier proteins have specific optical rotations that, on denaturation, become more negative. It is believed that this is due

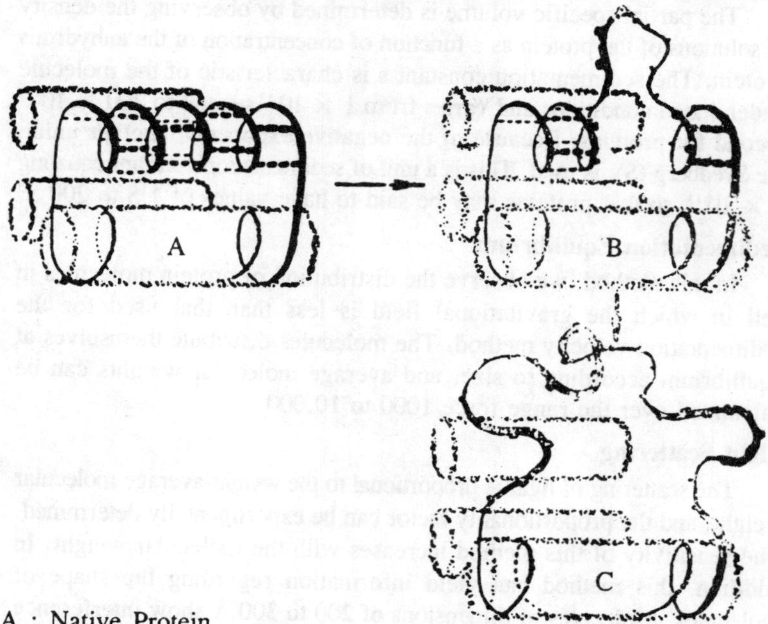

A : Native Protein

B : Initial Stage of Unfolding

C : Partially Denaturated Protein

Figure 5.11: Denaturation.

Table 5.6: Molecular Wieghts of Crystalline Proteins Determined by Various Methods

Method	*Ovalbumin*	*Serum Albumin*	*Hemoglobin*	*Lysozyme*	*p-Lactoglobulin*	*Insulin*
Containing weight	43,000	70,000	66,700	14,000	42,000	6000
Osmotic pressure	45,000-46,000	66,000-76,000	67,000	17,000	35,050-38,000	12,000-48,000
Diffusion	34,500		68,600	13,000	40,000	40,500
Sedimentation	43,800-45,000	65,360-70,200	68,000	14,000-17,000	38,000-41,500	12,000-48,000
Light scattering	45,700-46,000	72,500		14,700	33,700-35,500	12,000-48,000
X-radiation		65,600-66,200	66,700	13,900	33,000-35,600	36,000-52,400

to the disorganisation of the helical configuration, which lacks reflection symmetry and contributes, in the native protein, to the positive rotation, which keeps the specific rotation of the most native proteins at values of -30° to -60°.

On denaturation, the degree of rotation generally increases to values of -80° to -120°. Doty has used such measurements to estimate the helical content of proteins by assigning a value of $[\alpha]_D$=-90° for a 100 per cent random coil, and $[\alpha]_D$=-10° for a 100 per cent helix. The method of *optical rotatory*/dispersion—the variation of rotation with wave length of lighthas also been used to approximate the helical content of globular proteins. On changing from the native to the denatured state, the shape of the curve is changed markedly. Using this method it has been estimated that ribonuclease contains 15 per cent, whereas ovalbumin and serum albumin have a 45 per cent helical configuration. Tropomyosin has been estimated to exit in the helical form to an extent greater than 90 per cent.

DEUTERIUM EXCHANGE

Insight into protein structure may be obtained by determining the rate and extent of exchange of deuterium between the protein and the surrounding medium (water). Whereas the hydrogen in various groups of amino acids-e.g., amino carboxyl, amide, and hydroxyl-is instantaneously exchangeable with deuterium, the presence of strong internal hydrogen bonds in a protein may decrease this rate or eliminate it entirely. For example, at a pH of 4.7, the enzyme ribonuclease has some 243 exchangeable hydrogen atoms, of which 123 are peptide hydrogens.

In the native state, 185 atoms of hydrogen exchange promptly, whereas if the disulfide bonds are broken by oxidation or if the native enzyme is treated with a denaturing agent, all 243 atoms exchange instantaneously. It appears, therefore, that the hydrogen-bonded folded structure of the ribonuclease molecule shields some of the hydrogen atoms from instantaneous exchange.

TERTIARY STRUCTURE FROM X-RAY DIFFRACTION ANALYSIS: MYOGLOBIN AND HEMOGLOBIN

Just as Sanger was able to give us the first structural formula of a protein—insulin—Kendrew and his collaborators have presented us with first three-dimentional picture of a protein molecule reconstructed entirely from x-ray diffraction data. Myoglobin was the protein chosen for this study, since it has a molecular weight of only 17,000 and crystalises easily from muscle extracts of the sperm whale. It consists of a protein, globin (not the same as that found in hemoglobin), which is a single polypeplide of 153 amino acid residues, and one heme group. Heme (in oxymyoglobin) is composed of a substituted glutamic acid residue in Hb $M_{Milwaukee-1}$ leads to a stable complex. Again, this glutanic acid residues is four residues removed from his-63 and might therefore be in close approximation to the iron atom, since one turn of the helix accommodates the equivalent of 3.6 amino acid residues.

DENATURATION

Some denaturation phenomena suggest that tertiary structure is determined by amino acid sequence. Transformation from the native to the denatured state is often irreversible, although Anson and Mirsky, as early as 1930, demonstrated the reversibility of hemoglobin denaturation. Yet, if the native configuration is the most stable, denaturation should, under certain conditions, be reversible.

This reversibility has been demonstrated with the enzyme ribonuclease, which has four cystine disulfide bridges stabilising its tetiary structure. If ribonuclease is reduced to the —SH form, in the presence of the powerful denaturing agen urea, the enzyme loses all activity. Reoxidation of the reduced enzyme by air leads to a completely active enzyme indistinguishable from the native protein. With eight sulphydryl groups, the reduced enzyme might reoxidise in many ways. It would appear, therefore, that the amino acid sequence by itself is sufficient to determine the most stable configuration. It should be noted, however, that ribonuclease contains but one polypeptide chain. In the case of insulin which contains two chains, attempted oxidation of the reduced protein leads to a recovery of only 1 to 2 per cent of the original activity. It is possible that permanent damage is done or that bonds are formed during denaturation, which stabilises the unfolded or denatured, state. Many other enzymes have been reversibly denatured—e.g., catalase—to the extent of a 50 per cent yield.

We may summarise the various effects of denaturation

1. Decreased solubility at the isoelectric point.
2. Loss of biological activity.
3. Increased reactivity of chemical groups that had been masked by folding in the native protein—e.g., —SH, —S—S—, and phenolic hydroxyl groups of tyrosine.
4. Increased number of ionisable groups available for titration.
5. Increase in levorotation.

6. Increase in asymmetry of the molecule.
7. Increase in susceptibility to hydrolysis by proteolytic enzymes.
8. Loss of crystallizability.
9. Cleavage of hydrogen bonds.

QUATERNARY STRUCTURE

This level of organisation of the protein molecule results from the association of folded units into relatively stable aggregates. Many examples are available from molecular weight studies of the effect or change in environmental conditions on the degree of association or dissociation of proteins. For example, a 1 per cent solution of insulin yeilds a sedimentation constant of 3.7 S for the protein, corresponding to a molecular weight of 36,000. At acid pH a value of 2.0 S is obtained, which corresponds to a molecular weight of 12,000.

Some protein molecules are complexes of more than one polypeptide chain; each chain with its tertiary structure then constitutes a subunit of a larger molecule. The geometry of the combination of subunits into the complete molecule is said to be its quaternary structure.

The forces available for the formation of *quaternary structure* are the same as those that create secondary and tertiary structure. One of the most important is the tendency for hydrophobic groups to combine so as to exclude water. Suppose that a polypeptide chain folds in such a way as to create a face with exposed hydrophobic side chains. That face would tend to stick to any other hydrophobic surface.

If there are two polypeptide chain with areas of hydrophobic surface that can fit closely together, then the two chains will bond into a large molecule, and this is the way in which many proteins containing multiple subunits are constructed. (Of course, tertiary structure with an exposed hydrophobic face wouldn's be very stable as an isolated entity, but when two or more chains are present simultaneously that can form matching faces, otherwise transient areas of hydrophobic surface will be stabilised by combining with each other).

Liver alcohol dehydrogenase, for example, is a dimer that contains two identical subunits. The domains with pleated sheet cores that we discussed earlier comprise the tertiary structure of each subunit. However, these subunits have one segment of pleated sheet exposed at the surface in such a way that if two subunits about, a continuous pleated sheet will be formed from one subunit to the other. The subunits are then joined both by association of hydrophobic side chains and by hydrogen bonding.

Hemoglobin

Perhaps the most thoroughly studied example of quaternary structure is that of hemoglobin. The molecules of hemoglobin that carry oxygen in human blood are tetramers made from a pair of each of two kinds of polypeptide chains. These chains are designated α and β in the most abundant hemoglobin in adults, and this hemoglobin A therefore has the chain formula $\alpha_2\beta_2$.

Each of these chains has a tertiary structure much like that of myoglobin, which is made from only one polypeptide chain. However, they differ from myoglobin in that they are built to associate in particular ways. That is, α chains tend to bind β chains, and *vice versa*. The result is a tetrahedral molecule with the four subunits in close contact, except for a hole in the center that is freely accessible to water.

The hemoglobin chains associate and the myoglobin chains do not because some of the amino acid residues on the surfaces are different. The hemoglobin subunits have many additional hydrophobic residues exposed; they can only be buried by bringing the subunits together. They also have different polar side chains at locations where they form hydrogen bonds when the subunits about. It is the additional surface charge and the lack of hydrogen bonding that keep the myoglobin molecules separated. (The hemoglobin subunit surfaces are also contoured for closer contact).

On the other land, the surfaces remaining exposed after formation of the tetramer are designed to minimise any tendency for further association. Red blood cells are stuffed with hemoglobin almost to the limit of its solubility, with adjacent molecules not much more distant than they are in a crystal. Even a moderate tendency to aggregate would not permit such a high concentration, and therefore would diminish the capacity to carry oxygen. We shall that changes in single amino acid residues sometimes cause problems in this way.

What is the point of building hemoglobin, or any other protein, out of multiple polypeptide chains when many respectable protein have only one? Sometimes proteins are built this way simply to become large enough so. that they won't leak through membranes, but we shall see when we turn to the biological functions of proteins that those composed of more than one subunit can respond to change in environmental concentrations in a more complex way than can those made from only one poylpeptide chain. But before we examine these functions we should consider bow proteins are made, and this process involves another class of macromolecular polymers, the nucleic acids.

FUNCTION OF PROTEINS

Protein play crucial roles in virtually all biological processes. The significance and remarkable scope of their functions are exemplified in:

Enzymatic Catalysis

Nearly all chemical reactions in biological systems are catalysed by specific macromolecules called enzymes. Some of these reactions, such as the hydration of carbon dioxide, are quite simple. Others, such as the replication of an entire chromosome, are highly intricate. Nearly all enzymes exhibit enormous catalytic power. They usually enhance reaction rates by at least a millionfold. Indeed, chemical transformations reraly occur at perceptible rates in vivo in the absence of enzymes Several thousand enzymes have been characterised, and many of them have been crystallised. The striking fact is that all known enzymes are proteins. Thus, proteins play the unique role of determining the pattern of chemical transformations in biological systems.

Transport and Storage

Many small molecules and ions are transported by specific proteins. For example, hemoglobin tranports oxygen in erythrocytes, whereas myoglobin, a related protein, transports oxygen in muscle. Iron is carried in the plasma of blood by transferring and is stored in the liver as a complex with ferritin, a different protein.

Co-ordlnated Motion

Proteins are the major component of muscle. Muscle contraction is accomplished by the sliding motion of two kinds of protein filaments. On the microscopic scale, such co-ordinated motions as the movement of chromosomes in mitosis and the propulsion of sperm by their flagella also are produced by contractile assemblies consisting of proteins.

Mechanical Support

The high tensile strength of skin and bond is due to the presence of collagen, a fibrous protein.

Immune Protection

Antibodies are highly specific pr eins that recognise and combine with such foreign substances as viruses, bacteria, and cells from other organisms. Proteins thus play a vital role in distinguishing between self and nonself.

Generation and Transmission of Nerve Impulses

The response of nerve cells to specific stimuli is mediated by receptor proteins. For example, rhodopsin is the photoreceptor for protein in retinal rod cells. Receptor molecules that can be triggered by specific small molecules, such as acetlycholine, are responsible for transmitting nerve impulses at synapsesthat is, at junctions beween nerve cells.

Control of Growth and Differentiation

Controlled sequential expression of genetic information is essential for the orderly growth and differentiation of cells. Only a small fraction of the genome of a cell is expressed at any one time. In bacteria, repressor proteins are important control elements that silence specific segments of the DNA of a cell. A quite different way in which proteins act in differentiation is exemplified by nerve growth factor, a protein complex that guides the formation of neutral networks in higher organisms.

6

ENZYMES AND BIOLOGICAL CATALYSIS

A living organism is a magnificent assembly of chemical reactions, an entity rather than chaos because most of its components make an orderly contribution to existence of the whole. Many of these reactions occur at significant rates only because specific proteins—the enzymes—are present to catalyse them. The enzymes are constructed to control both the kinds of reaction and the rates at which they occur, and it is the existence of these proteins that creates a complex harmony of chemical function. Enzymes contain the same amino acids found in other proteins, and they have a three-dimensional structure that represents the conformation of least energy content, as do other proteins.

All of our present information supports the view that enzymatic catalysis involves the same kinds of bonds that appear during ordinary organic reactions, using structures supplied by amino acid residues and by prosthetic groups. The binding of prosthetic groups is itself determined by conformation of amino acid residues, so the study of the nature of enzymes is in many ways only an extension of the principles of protein structure that we have already developed.

There are thousands and thousands of different kinds of enzymes, each designed to enhance the rate of a specific reaction. As a result of a century of intensive laboratory investigation, much is now known about what enzymes are, how they work, and how they regulated. Beause of the ubiquitous role of enzymes in cell, it is important for us to consider these general issues this early stage in the book.

DISCOVERY OF BIOLOGICAL CATALYSTS

In the early nineteenth century the Swedish chemist Jon Berzelius discovered that an extract of potatoes is more effective than concentrated sulphuric acid in promoting the breakdown of starch. Berzelius concluded that this effect is due to the presence of biological catalysts and with remarkable insight went on to predict that all materials found in living organisms are synthesized under the influence of such catalysts.

This observation stimulated interest in the biochemical identity of these catalysts, and Louis Pasteur soon postulated that the catalytic effect is intimately associated with cell structure and so cannot be separated from living cells. If true, this meant that the catalysts could not be isolated and studied in purified form.

It was therefore a great milestone when, in 1897, Edward Buchner demonstrated that an extract of yeast, from which all intact cells bad been removed, is capable of catalyzirg the breakdown of glucose. This was an especially significant observation because glucose degradation is a major metabolic pathway in cells. This finding firmly established that biological catalysts, in the absence of any cell organisation, are capable of enhancing the rates of metabolic reactions.

The increasing importance of this class of molecules soon led to suggestions for a uniform nomenclature system to facilitate communication among biochemists. The term *enzyme* was adopted as a general designation for biological catalysts, and it was decided that individual enzymes would be named by adding the suffix "*ase*" to the name of the *substrate* (the substance upon which the enzyme acts). For example, the enzyme catalysing the degradation of urea was termed urease, enzymes catalysing hydrolysis of nucleic acids were designated nucleases, the enzyme catalysing glucose oxidation was called glucose oxidase, and so forth.

This basic approach to enzyme nomenclature is still in use today, although some older names introduced prior to the adoption of this system (e.g., trypsin, pepsin) have been retained. In the years immediates following the initial demonstration of cell-free enzyme activity by Buchner, a large number of other enzymes were identified by biochemists. But little progress was made in understanding the structure or mechanism of these cellular catalysts.

An example of the difficulties experienced during this period is the classic story of Richard Willstatter's encounters with the enzyme *peroxidase,* one of the first enzymes to be obtained in a highly purified form. When Wilistatter chemically analysed his purified enzyme preparation, he found no lipid, no sugar (and therefore no nucleic acid), and no protein. These results led to the prevailing opinion in the early twentieth century that enzymes belong to none of the commonly recognised classes of biological molecules. In 1926, however, James Sumner reported the first crystallisation of an enzyme, in this case urease. He found that the crystals consisted of pure protein, even after repeated cycles of recrystallisation.

Since crystallisation is generally recognised as the ultimate criterion of purity, these results suggested to Summer that the enzyme urease is a protein. Although this conclusion initially met with skepticism, similar experiments by John Northrop and his colleagues on other enzymes soon led to general acceptance of the idea that enzymes are made of protein. Willstatter, it turns out, had been misled by the fact that enzymes are active in extremely small quantities.

Thus his peroxidase preparation exhibited measurable enzyme activity even though the protein content of the sample was too low to be detected by the analytical techniques then available. In the years since the pioneering work of Sumner and Northrop, over a thousand enzymes have been purified and identified as proteins. Though the number of different reactions catalysed by these enzymes is very large, one can subdivide this group into a relatively small number of categories.

All cellular reactions are catalysed by enzymes that fit into one of these categories. As Berzelius predicted many years ago, enzymes act as extraordinarily efficient chemical catalysts. Enzymes therefore exhibit the four defining characteristics shared by all catalysts:

(1) They modify the rates of chemical reactions.

(2) They are not consumed in the process.

(3) They are effective in extremely small quantities.

(4) They do not alter the equilibrium of the reaction being catalysed.

In addition to these four attributes common to all catalysts, enzymes exhibit several unique features that distinguish them from the catalysts routinely encountered in organic or inorganic chemistry. First, enzymes are exceedingly efficient compared to other catalysts. A typical enzymatic reaction may proceed as much as 10^8-10^{11} times faster than the same reaction catalysed by a nonenzymatic catalyst. Also, unlike the nonbiological catalysts routinely encountered in chemistry, the catalytic activity of enzymes can be regulated to suit specific cellular needs. Finally, one of the most striking properties of enzymes is their selectivity. Unlike other catalysts enzymes are highly specific, both in the types of

Table 3.1: Current nomenclature of six types of biochemical reactions catalysed by enzymes.

Enzyme Class	*Type of Reaction Catalysed*	*Example*
Oxidoreductases	Oxidation-reduction	Dehydrogenases, oxidases, perioxidases, hydroxylases, reductases, oxygenases
Transferases	Transfer of a functional group from one molelases cule to another (AX+B → A+BX)	Transaminases, transmethylases
Hydrolases	Hydrolytic cleavage	Esterases, amidases glycosidases, peptidases, phosphatases
Lyases	Cleavage of C—C, C—O, C—N, etc. by elimination; leaving a double bond ; or addition to double bonds	Aldolases, synthases, deaminases, hydrases, decarboxylases, nucleotide cyclases
Isomerases	Intramolecular rearrangements	Isomerases; racemases, epimerases, mutases
Ligases	Joining together of five molecules coupled to the cleavege of a highenergy bond	Synthetases

reactions they catalyse and in the structures of the substances upon which they act. For example, the enzyme glucose oxidase is specific not just for the oxidation of glucose versus other six-carbon sugars but also for a particular three-dimensional form of the sugar, namely, α-glucose.

The unique properties of enzymes (efficiency, regulation, and specificity) all stem from the fact that enzymes are protein molecules and therefore exhibit all the remarkable characteristics of proteins. They study of how enzymes work is therefore intimately tied to the study of protein chemistry.

Catalytic Site

Why then do some proteins act as very effective catalysts-more effective than any other known compounds in making some reactions go faster ? The answer lies in the ability of proteins to be molded into a variety of shapes. Proteins acting as enzymes are built with clefts that function as catalytic sites. Many compounds can wander in and out, but when one with a proper fit diffuses in, the protein closes around it and brings to bear chemical groups in a particular geometric conformation.

Here is the basis for enzyme function: The catalytic effectiveness of amino, carboxyl, and other bonding groups is increased by several order of magnitude placing them in fixed spatial arrangements where they can lock onto the affected compound; yet it is possible to stretch or otherwise deform particular bonds through relatively small changes in conformation of the protein.

We shall see that the use of proteins as catalysts conveys other advantages. They can be constructed with separate domains, making rapid multimolecular reactions feasible in some cases, or providing binding sites for allosteric effectors with which to regulate the rates of reactions.

Specificity

The *catalytic she* of enzymes is a part of a *binding site*. Enzymes, like antibodies, are frequently highly specific in binding particular compounds, because the amino acid residues in the binding site closely fit only a few compounds, stronger force with "good" substrates than with other compounds. The existence of this stronger force is an important factor in easy deformation of the substrate to break old bonds and create new ones. Enzymes differ from antibodies in that tight binding may change the bound compound.

Catalytic Groups

Another clue to the catalytic mechanism comes from the presence in proteins of groups known to be effective catalysts for ordinary organic reactions. For example, many amino acid side chains are capable of donating or accepting protons so as to act as *general acid* or *general base catalysts*. Other groups, such as hydroxyl or amino groups, may act as *nucleophiles*, and surrounding structures in an enzyme are often designed to reinforce their catalytic properties. (Nucleophiles in effect donate electrons to bond other groups lacking filled electron orbitals.)

General Catalytic Mechanism

Enzymes owe their peculiar catalytic effectiveness to a combination of specific binding and the presence of catalytic groups. In an ordinary chemical reaction, a compound must confort into an unfavourable configuration before it changes into something else the formation of the less probable intermediate state represents an activation energy barrier that must be overcome before reaction can occur. Enzymes change this situation by being built so that their combination with the *reacting* compound is most Stable when the compound is in a "*high-energy*" intermediate form.

An enzymatic reaction proceeds through at least three stages the formation of a complex between enzyme and substrate, the conversion of this complex to an enzy me-intermediate complex, and the further conversion to a complex between enzyme and product that can dissociate

$$E + S \rightleftharpoons ES \rightleftharpoons EI \rightleftharpoons EP \rightleftharpoons E + P$$

The groups on the enzyme are arranged to make the total energy of the enzyme intermediate complex less than the total energy in the enzyme-substrate or enzyme-product complexes. The activation barriers in converting enzyme-substrate to enzyme-intermediate, or enzyme-intermediate to enzyme-product are substantially lower than the barrier without enzyme.

$$\begin{array}{ccl} E & \overset{+S}{\rightleftharpoons} & ES \rightarrow E+P \\ +1 \upharpoonleft\downharpoonright & & \upharpoonleft\downharpoonright +1 \\ & \underset{+S}{\rightleftharpoons} & ESI \end{array}$$

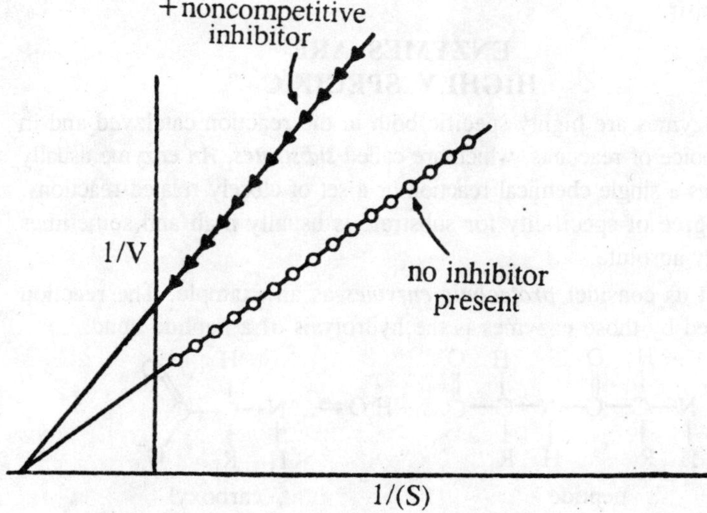

Figure 3.1: A double-reciprocal plot of enzyme kinetics in the presence (Δ-Δ-Δ) and absence (O-O-O) of a noncompetitive inhibitor ; K_M is unaltered by the non-competitive inhibitor whereas V_{max} is decreased.

The specific binding of substrate to enzyme therefore serves not only to confine particular reactions to a few compounds but also to create a complex that spontaneously favours the formation of a more reactive configuration. The enzyme does several things to achieve this end.

It provides catalytic groups in a single molecule, eliminating the need for collision with two or more molecules during the course of the reaction (*entropy effect*); it aligns the substrate at favourable angles relative to the catalytic groups so that the interaction with these groups is facilitated *(orbital steering)*, and it provides binding groups at Positions that stabilise a reactive intermediate (*propinquity effect*).

Perhaps the easiest way to get a feel for what enzymes do and how they go about it is to examine a few of them in detail, not because these details are something that ought to be committed to memory, but so we can develop some general principles by example.

Enzyme mechanisms are rarely "*seen*" in the sense that they can be frozen for inspection by x-ray crystallography ; they are *deduced* through the use of techniques such as molecular probes that react with active sites, destruction of selected amino acids, spectral changes, and more recently by neutron diffraction ; hence our concentration on the principal features rather than the more conjectural fine points of mechanism.

ENZYMES ARE HIGHLY SPECIFIC

Enzymes are highly specific both in the reaction catalzyed and in their choice of reactants, which are called *substrates. An* enzyme usually catalyzes a single chemical reaction or a set of closely related reactions. The degree of specificity for substrate is usually high and sometimes virtually absolute.

Let us consider *proteolytic enzymes* as an example. The reaction catalysed by those enzymes is the hydrolysis of a peptide bond.

$$\underset{\text{peptide}}{-\mathrm{N(H)}-\mathrm{C(H)(R_1)}-\mathrm{C(=O)}-\mathrm{N(H)}-\mathrm{C(H)(R_2)}-\mathrm{C(=O)}\ldots} + H_2O \rightleftharpoons \underset{\text{carboxyl competent}}{\ldots\mathrm{N(H)}-\mathrm{C(H)(R_1)}-\mathrm{C(=O)O^-}}$$

peptide

carboxyl competent

Most proteolytic enzymes also catalyse a different but related reaction, namely the hydrolysis of an ester bond.

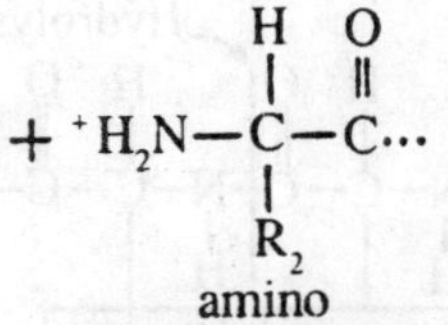

$$+\ {}^{+}H_2N-\overset{\overset{\displaystyle H}{|}}{\underset{\underset{\displaystyle R_2}{|}}{C}}-\overset{\overset{\displaystyle O}{\|}}{C}\cdots$$

amino component

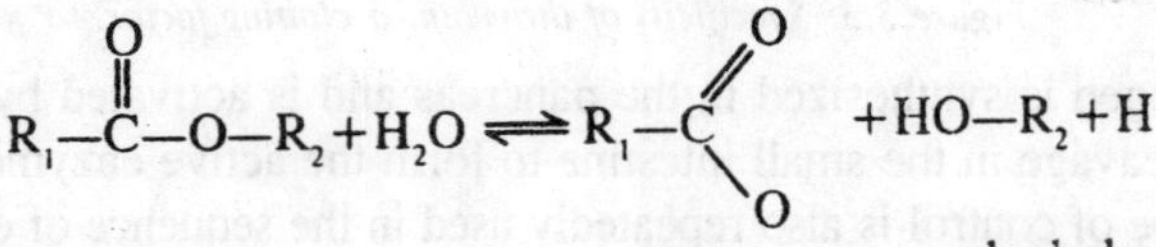

$$R_1-\overset{\overset{\displaystyle O}{\|}}{C}-O-R_2 + H_2O \rightleftharpoons R_1-C\begin{matrix}\nearrow O \\ \searrow O\end{matrix} \quad + HO-R_2 + H^{+}$$

ester acid alcohol

Proteolytic enzymes vary markedly in their degree of substrate specificity. Subtilisin, which comes from certain bacteria, is quite undiscriminating about the nature of the side chains adjacent to the peptide bond to be cleaved. Trypsin, as was mentioned in previous Chapter is quite specific in that it splits peptides bond on the carboxyl side of lysine and residues only. Thrombin, an enzyme participating in blood clotting, is even more specific than trypsin. The side chain on the carboxyl side of the susceptible peptible bond must be arginine, whereas the one on the amino side must be glycine.

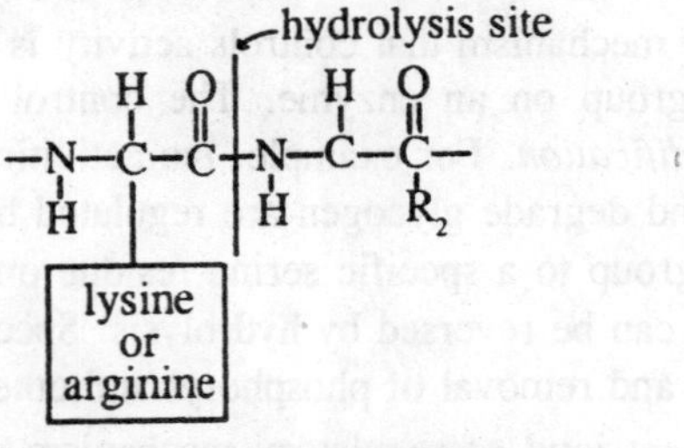

hydrolysis site

$$-\underset{\underset{\displaystyle H}{|}}{N}-\overset{\overset{\displaystyle H}{|}}{C}-\overset{\overset{\displaystyle O}{\|}}{C}-\underset{\underset{\displaystyle H}{|}}{N}-\overset{\overset{\displaystyle H}{|}}{C}-\underset{\underset{\displaystyle R_2}{|}}{\overset{\overset{\displaystyle O}{\|}}{C}}-$$

lysine or arginine

Figure 3.2 : Specificity of trypsin.

Another example of the high degree of specificity of enzymes is provided by DNA polymerase 1. This enzyme synthesizes DNA by linking together four kinds of nucleotide building blocks. The sequence of nucleotides in the DNA strand that is being synthesized is determined by the sequence of nucleotides in another DNA strand that serves as a template DNA polymerase I is remarkably precise in carrying out the instructions given by the template. The wrong nucleotide is inserted into a new DNA strand less than once in a million times.

Some enzymes are synthesized in an *inactive precursor form* and are activated at a physiologically appropriate time and place. The digestive enzymes exemplify this kind of control. For example,

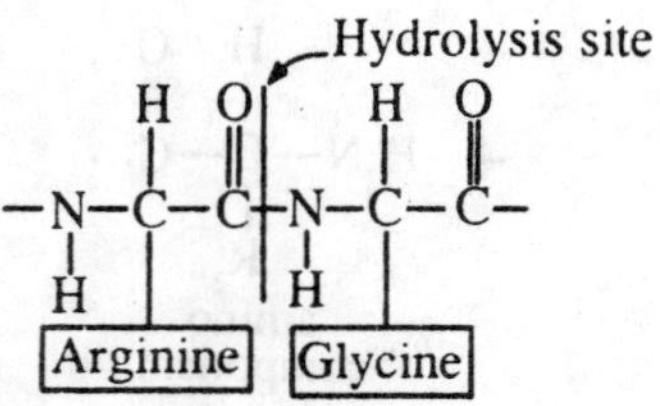

Figure 3.3: Specificity of thrombin, a clotting factor.

trypsinogen is synthesized in the pancreas and is activated by peptide-bond cleavage in the small intestine to form the active enzyme trypsin. This type of control is also repeatedly used in the sequence of enzymatic reactions leading to the clotting of blood. The enzymatically inactive precursors of proteolytic enzymes are called *zymogens*.

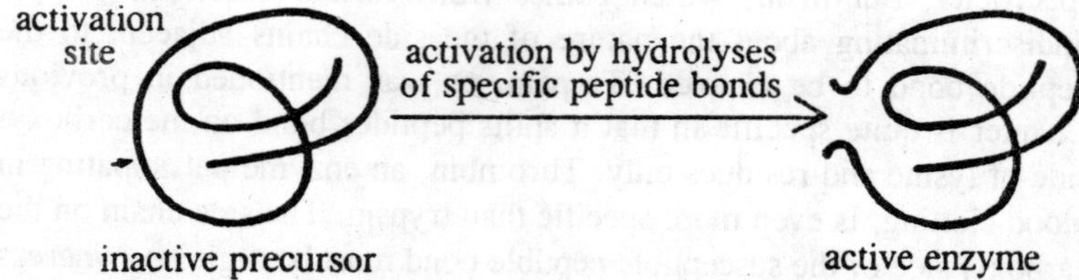

Figure 3.4: Zymogen activation by hydrolysis of specific peptide bonds.

Another mechanism that controls activity is the covalent insertion of a small group on an enzyme. The control mechanism is called *covalent modification*. For example, the activities of the enzymes that synthesize and degrade glycogen are regulated by the attachment of a phosphoryl group to a specific serine residue on these enzymes. This modification can be reversed by hydrolysis. Specific enzymes catalyze the insertion and removal of phosphoryl and other modifying groups.

A different kind of regulatory mechanism affects many reaction sequences resulting in the synthesis of small molecules such as amino acids. The enzyme that catalyses the first step in such a biosynthetic pathway is inhibited by the ultimate product. The biosynthesis of isoleucine in bacteria illustrates this type of control, which is called *feedback inhibition*. Threonine is converted into isoleucine in five steps, the first of which is catalysed by threonine deaminase.

This enzyme is inhibited when the concentration of isoleucine reaches a sufficiently high level. Isoleucine binds to a regulatory site on the enzyme, which is distinct from its catalytic site. The inhibition of threonine deaminase is mediated by an *allosteric interaction*, which is reversible. When the level of isoleucine drops sufficiently, threonine deaminase becomes active again, and consequently isoleucine is again synthesized.

The specificity of some enzymes is under physiological control. The synthesis of lactose by the mammary glands is a particularly striking example. Lactose, synthetase, the enzyme that catalyzes the synthesis of lactose, consists of a catalytic subunit and a modifier subunit. The catalytic subunit by itself cannot synthesize lactose.

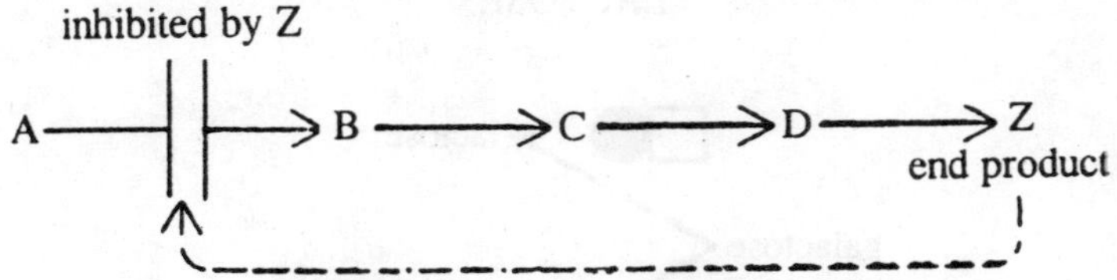

inhibited by Z

A ——||——→ B ——→ C ——→ D ——→ Z

end product

Figure 3.5 : Feedback inhibition of the first enzyme in a pathway by reversible binding of the final product.

It has a different role, which is to catalyze the attachment of galactose to a protein that contains a covalently linked carbohydrate chain. The modifier subunit alters the specificity of the catalytic subunit so that it links galactose to glucose to form lactose. The level of the modifier subunit under hormonal control. During pregnancy, the catalytic subunit is formed in the mammary gland but little modifier subunit is formed.

At the time of birth, hormonal levels change drastically, and the modifier subunit is synthesized in large amounts. The modifier subunit then binds to the catalytic subunit to form an active lactose synthetase complex that produces large amounts of lactose. This system clearly shows that hormones can exert their physiological effects by altering the specificity of enzymes.

ENZYMES TRANSFORM DIFFERENT KINDS OF ENERGY

In many biochemical reactions, the energy of the reactants is converted into a different form with high efficiency. For example, in photosynthesis, light energy is converted into chemical-bond energy. In mitochondria, the free energy contained in small molecules derived from foods is converted into a different currency, that of adenosine triphosphate (ATP). The chemical-bond energy of ATP is then utilised in many different ways.

In muscular contraction, the energy of ATP is converted into mechanical energy. Cells and organelles have pumps that utilise ATP to transport molecules and ions against chemical and electrical gradients.

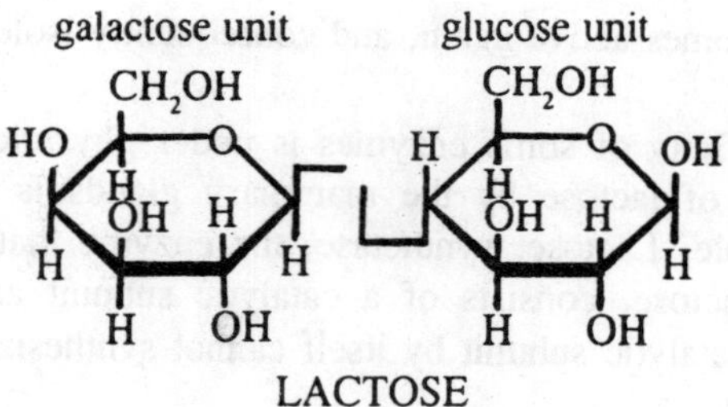

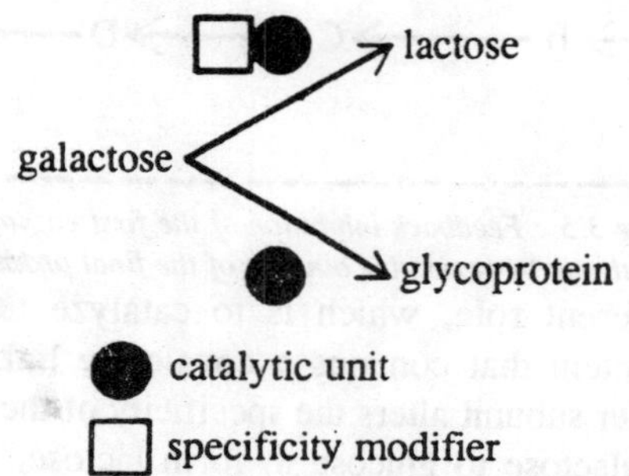

Figure 3.6: Lactose, a sugar consisting of a galactose and a glucose residue, is synthesised by an enzyme that contains a catalytic subunit and a specificity-modifier subunit. A different reaction is catalysed by the catalytic subunit alone.

These transformations of energy are carried out by enzyme molecules that are integral parts of highly organised assemblies.

ENZYMES DO NOT ALTER REACTION EQUILIBRIE

An enzyme is a catalyst and consequently it cannot a the equilibrium of a chemical reaction. This means that enzyme accelerates the forward and reverse reaction by precisely the same factor.

Consider the interconversion of A and B. Suppose that in the absence of enzyme the forward rate (kF) is 10^{-4} sec^{-1} and the reverse rate (k_R) is 10^{-8} sec^{-1}. The equilibrium constant K is given by the ratio of these rates

$$A \underset{10^{-6}\ sec^{-1}}{\overset{10^{-4}\ sec^{-1}}{\rightleftharpoons}} B$$

$$K = \frac{[B]}{[A]} = \frac{k_F}{k_R} = \frac{10^{-4}}{10^{-6}} = 100$$

The equilibrium concentration of B is 100 times that of A, whether or nDt enzyme is present. However, it would take several hours to approach this equilibrium without enzyme, whereas equilibrium would be attained within a second when enzyme is present. Thus, enzymes accelerate the attainment of equilibria but do not shift their positions.

ENZYMES DECREASE THE ACTIVATION ENERGIES OF REACTIONS CATALYSED BY THEM

A chemical reaction, AvaB, goes through a transition *state* that has a higher energy than either A or B. The rate of the forward reaction depends on the temperature and on the difference in free energy between that of A and the transition state, which is called the *Gibbs free energy of activation* and symbolised by $\Delta G_{\dagger}$.

$$\Delta G_{\dagger} = G_{\text{transition state}} - G_{\text{substrate}}$$

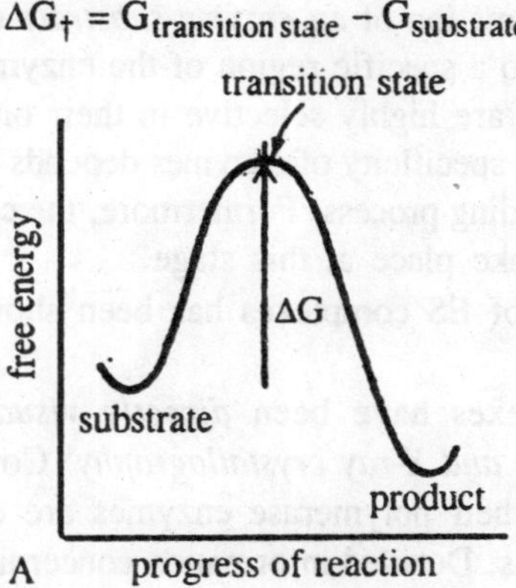

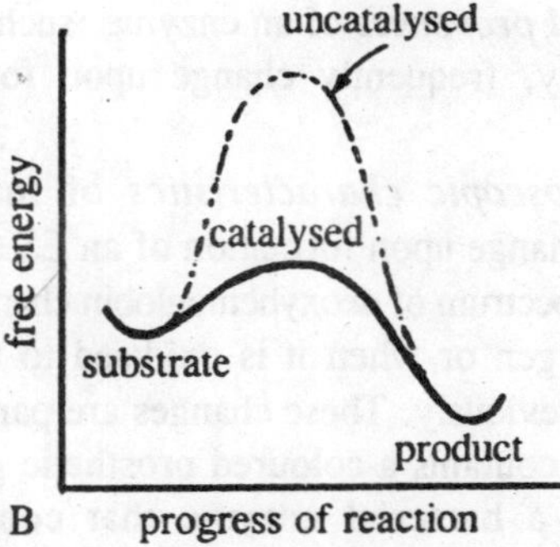

Figure 3.7: A. Definition of $\Delta G_{\dagger}$, the free energy of activation. B. Enzymes accelerate catalysis by reducing $\Delta G_{\dagger}$.

The reaction rate is proportional to the fraction of molecules that have a free energy equal to or greater than AG$. The proportion of molecules that have an energy equal to or greater than AG$ increases with temperature.

Enzymes accelerate reactions by decreasing $\Delta G_{\dagger}$, the activation barrier. The combination of substrate and enzyme create. a new reaction pathway whose transition-state energy is lower than it would be if the reaction were taking place in the absence of enzyme.

FORMATION OF AN ENZYME-SUBSTRATE COMPLEX IS THE FIRST STEP IN ENZYMATIC CATALYSIS

The making and breaking of chemical bonds by an enzyme are preceded by the formation of an *enzyme-substrote* (ES) complex. The substrate is bound to a specific region of the enzyme called the *active site*. Most enzymes are highly selective in their binding of substrate. Indeed, the catalytic specificity of enzymes depends in large part on the specificity of the binding process. Furthermore, the control of enzymatic activity may also take place at this stage.

The existence of ES complexes has been shown in a variety of ways:

1. ES complexes have been *directly visualised by electron microscopy and X-ray crystallography*. Complexes of nucleic acids and their polymerase enzymes are evident in electron micrographs. Detailed information concerning the location and interactions of glycyl-L-tyrosine, a substrate of carboxypeptidase A, has been obtaincd from X-ray studies of that ES complex.
2. The *physical properties* of an enzyme, such as its solubility or heat stability, frequently change upon formation of an ES complex.
3. The *spectroscopic characteristics* of many enzymes and substrates change upon formation of an ES complex just as the absorption spectrum of deoxyhemoglobin changes markedly when it binds oxygen or when it is oxidised to the ferric state, as described previously. These changes are particularly striking if the enzyme contains a coloured prosthetic group. Tryptophan synthetase, a bacterial enzyme that contains a pyridoxal phosphate prosthetic group, affords a nine illustration. This enzyme catalyzes the synthesis of L-tryptophan from L-serine and indole. The addition of L-serine to the enzyme produces a

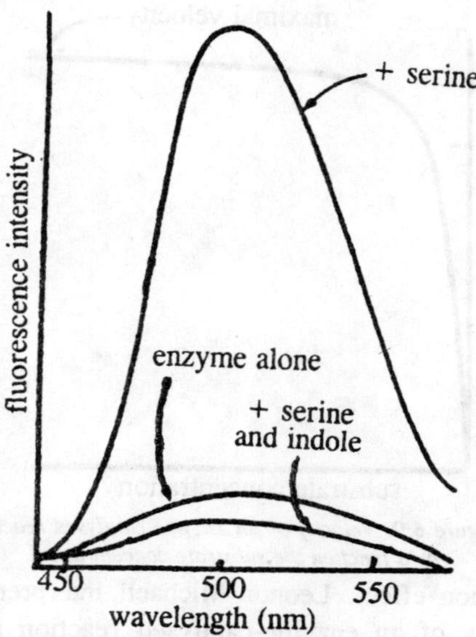

***Figure 6.8:** Fluorescence intensity of the pyridoxal phosphate group at the active site of tryp ophan synthetase changes upon addition of serine and indole, the substrates.*

marked increase in the fluorescence of thepyridoxal phosphate group. The subsequent addition of indole. the second substrate, quenches this fluorescence to a level lower than that of the enzyme alone. Thus, fluorescence spectroscopy reveals the existence of an enzyme-serine complex and of an enzyme-serine indole complex. Other spectroscopic techniques, such as nuclear and electron magnetic resonance, also are highly information about ES interactions.

4. A high degree of *stereospecifieity* is displayed in theformation of ES complexes. For example, D-serine is not a substrate of tryptophan synthetase. Indeed, the D-isomer does not even bind to the enzyme. This implies that the substratebinding site has a very well define shape.
5. ES complexes can sometimes be *isolated in pure from.* For an enzyme that catalyzes the reaction $A+B \rightleftharpoons C$, it is sometimes possible to isolate an EA complex. This can be done the enzyme has a sufficiently high affinity for A and if B absent from the mixture.
6. At a constant concentration of enzyme, the reaction rat increases with increasing substrate concentration until maximal velocity is reached. In contrast, uncatalysed reaction do not show this

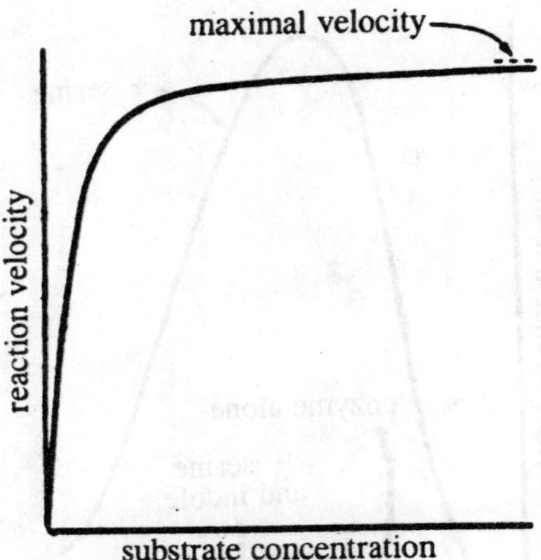

Figure 6.9: Velocity of an enzyme-catalysed reaction as a function the substrate concentration.

saturation effect. Leonor Michaeli interpreted the maximal velocity of an enzyme-catalysed reaction in terms of the formation of a discrete ES complex. At a sufficiently high substrate concentration, the catalytic sites are filled and so the reaction rate reaches a maximum. This is the oldest and most general evidence for the existence of ES complexes.

SOME FEATURES OF ACTIVE SITES

The active site of an enzyme is the region that binds the substrates (and the prosthetic group, if any) and contributes the residues that directly participate in the making and breaking of bonds. These residues are called the *catalytic groups*. Although enzymes differ widely in structure, specificity, and mode of catalysis, a number of generalisations concerning their active sites can he stated:

1. *The active site takes up a relatively small part of the total volume of an enzyme.*

Most of the amino acid residues in an enzyme are not in contact with the substrate. This raises the intriguing question of why enzymes are so big. Nearly all enzymes are made up of more than 100 amino acid residues, which gives them a mass greater than 10 kdal and a diameter of more than 25 Å.

2. *The active site is a three-dimensional entity*

The active site of an enzyme is not a point, a line, or even a plane. It is an intricate three-dimensional form made up of groups that come from different parts of the linear amino acid sequence-indeed, residues far apart in the linear sequence may interact more strongly than adjacent residues in the amino acid sequence, as has already been seen for myoglobin and hemoglobin. In lysozyme, an enzyme that will be discussed in more detail in the next chapter, the important groups in the active site are contributed by residues numbered 35, 52, 62, 63 and 101 in the linear sequence of 129 amino acids.

3. *Substrates are bound to enzymes by relatively weak forces*

ES complexes usually have equilibrium constants that range from 10^{-2} to 10^{-8} M, corresponding to free energies of interaction ranging from –3 to –12 kcal/mol. These values should be compared with the strengths of covalent bonds, which are between –50 and 110 kcal/mol.

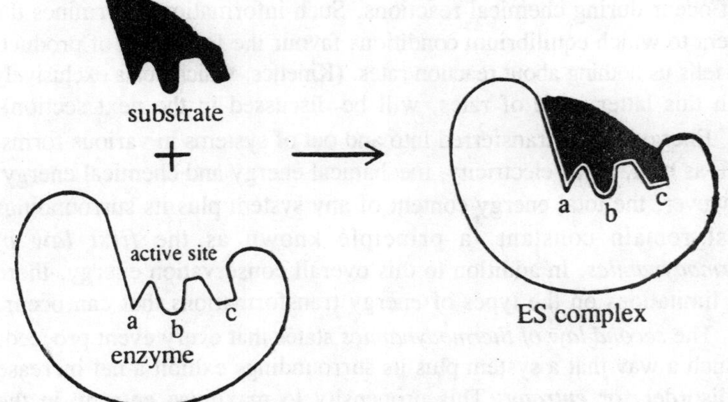

Figure 6.10 : Lock and-key model of the interaction of substrates and enzymes. The active site of the enzyme by itself is complementary in shape to that of the substrate.

4. *Active sites are clefts or crevices*

In all enzymes of known structure, substrate molecules are bound to a cleft or crevice from which water is usually excluded unless it is a reactant. The cleft also contains several polar residues that are essential for binding and catalysis. The nonpolar character of the cleft enhances the binding of substrate. In addition, the cleft creates a microenvironment in which certain polar residues acquire special properties essential for then catalytic role.

5. *The specificity of binding depends on the precisely defined arrangement of atoms in on active site*

A substrate must have a matching shape to fit into the site. Emil Fischer's metaphor of the lock and key, stated in 1890, has proved to be an essentially correct and highly fruitful way of looking at the stereospecificity of catalysis. However, recent work suggests that the active sites of some enzymes are not rigid. In such an enzyme, the shape of the active site is modified by the binding of substrate. The active has a shape complementary to that of the substrate only after the substrate is bound. This process of dynamic recognition is called *induced fir*. Further, some enzymes preferentially bind a strained form of the substrate corresponding to the transition state.

THERMODYNAMICS OF CHEMICAL REACTIONS

In order to discuss the question of how enzymes work, one must first have a basic familiarity with the concepts. of thermodynamics and kinetics. *Thermodynamics* refers to the study of the changes in energy that occur during chemical reactions. Such information determines the extent to which equilibrium conditions favour the formation of product, but tells us nothing about reaction rates. (Kinetics, which deals exclusively with this latter issue of rates, will be discussed in the next section).

Energy can be transferred into and out of systems in various forms, such as heat, light, electricity, mechanical energy and chemical energy. However, the total energy content of any system plus its surroundings must remain constant, a principle known as the *first law of thermodynamics*. In addition to this overall conservation energy, there are limitations on the types of energy transformations that can occur.

The *second law of thermodynamics* states that every event proceeds in such a way that a system plus its surroundings exhibit a net increase in disorder, or *entropy*. This propensity to maximise entropy in the universe is the ultimate driving force of all chemical reactions. The portion of the total energy used to increase the entropy is not available to do work. This amount of energy can be calculated by multiplying the change in entropy (LS) times the absolute temperature (T), yielding TLS. The energy available to do work, on the other hand, is known as the *free energy* (G). The changes in free energy and entropy occurring in a system undergoing transformation are related by the equation

$$\Delta G + L\Delta S = \Delta H$$

where H equals the *enthalpy* or beat content. In biological' systems, where pressure and volume are constant, enthalpy is equivalent to the total energy. Hence the above equation, which combines both the first and second laws of thermodynamics, tells us that in any system undergoing transformation, the total change in energy is equal to the changes in (1) free energy and (2) the energy involved in altering the entropy.

Although the second law of thermodynamics tells us that the entropy of a system plus its surroundings always proceeds toward a maximum, this does not necessarily mean that the entropy of an individual system always increases. It may either increase, decrease, or stay the same. If the system decreases in entropy, however, the entropy of the surroundings must increase by a sufficient amount so that the total entropy of system plus surroundings increases. This is exactly what happens during the development of living organisms. As highly organised living cells are produced, there is a decrease in entropy of the system, but only at the expense of an even greater increase in entropy of the surroundings.

The fact that a net increase in entropy accompanies all events proceeding toward equilibrium means that there must be a comparable decrease in free energy (due in essence to the loss of the free energy used to increase the entropy.) Because of this tendency to frec energy to decrease to a minimum, freeenergy calculations are useful indicators of the direction in which reactions tend to proceed. They are especially useful in this regard because changes in free energy are much easier to determine than changes in entropy.

Determination of Free-Energy Changes

For the generalised chemical reaction

Reactants → products

the change in free energy (ΔG) can be calculated from the equation

$$\Delta G = \Delta G° + 2.30RT \log_{10} \frac{[\text{products}]}{[\text{reactants}]} \qquad (6.1)$$

where R is the gas contants, T is the absolute temperature, [reactants] is the mathematical product obtained by multiplying together the initial molar concentrations of each of the reactants, [products] is the mathematical product obtained by multiplying together the initial molar concentrations of each of the products of the reaction, and ΔG° is the standard freeenergy change. This *standard free-energy* change (ΔG°) is a measure of the amount of free energy released during conversion of reactants to products under "*standard conditions*" (defined as all reactants and products present at an initial concentration of 1.0 M). It is critical that the distinction between ΔG and ΔG° be clearly understood. On the one hand ΔG° is a measure of the *actual* free-energy change that occurs given a mixture of reactants and products at any particular concentrations; the value of ΔG° thus varies, depending on the concentrations involved. On the other hand ΔG° is a constant for any given reaction at a given temperature. It is the free energy change that occurs under standard conditions of reactant and product concentration.

When equilibrium is achieved, it means by definition that no further net change in free energy can occur, that is, ΔG=O. Substituting ΔG±O into Equation 2.1, we obtain the equilibrium expression

$$0 = \Delta G° + 2.303RT \log_{10} \frac{[\text{product}_{eq}]}{[\text{reactants}_{eq}]}$$

where [reactants_{eq}] and [products_{eq}] are the molar concentrations of reactants and products at equilibrium, respectively. Rearranging terms we obtain

$$\Delta G° = 2.303RT \log_{10} \frac{[\text{products}]}{[\text{reactants}]} \qquad (6.2)$$

Because the ***equilibrium constant*** (K'_{eq}) for any chemical reaction is defined as

$$K'_{eq} = \frac{[\text{products}]}{[\text{reactants}]} \qquad (6.3)$$

K'_{eq} can be substituted into Equation 2.2 to obtain the general relationship

$$\Delta G° = -2.303RT \log_{10} K'_{eq} \qquad (6.4)$$

This final equation provides us with a relatively easy way of determining the standard free-energy change for any particular chemical reaction. One simply measures the concentrations of reactants and products after equilibrium has been achieved, uses this information to calculate K'_{eq} (Equation 2.3), and then substitutes this value of K'_{eq} into Equation 2.4 obtain ΔG°.

It can be readily calculated from Equation 2.4 that reactions with an equilibrium constant of 1.0 (equal concentrations of reactants and products at equilibrium) exhibit aΔG° of 0. This means that no free-energy change occurs during conversion of reactants to products under standard conditions. For reactions with an equilibrium constant greater than 1.0 (equilibrium favouring products). ΔG° will be negative and free energy will therefore be released during conversion of reactants to products. When the equilibrium constant is less than 1.0 (equilibrium favouring reactants), ΔG° will be positive and free energy will therefore be consumed during conversion of reactants to products.

Reactions with a negative ΔG°, termed *exergonic* reactions, can produced spontaneously in the direction written. However, it is important to emphasize that in this particular context, the term *spontaneous* does not signify anything about the speed at which the reaction will occur. A spontaneous reaction may take seconds or years to achieve equilibrium; its spontaneity refers only to the fact that no net input of energy is required in order to make the reaction proceed. Reactions in which $\Delta G°$ is positive are termed *endergonic,* and cannot proceed spontaneously in the direction written. Instead, they will proceed in the reverse direction.

In a living cell, of course, one does not generally start with 1.0 M concentrations of all ingredients, the standard conditions that apply to the above discussion of $\Delta G°$. Under other conditions it is ΔG, the actual free-energy change, that determines the direction in which a reaction proceeds. Depending on the particular conditions involved, ΔG may be smaller, larger, or the same as $\Delta G°$.

As an example to help summarise the above points, let us consider the following reaction that occurs as part of the pathway for metabolising glucose

CH_2OH | $C{=}O$ | $CH_2OPO_3^{2-}$	$\rightleftharpoons$	O H \\\\ / C | H—C—OH | $CH_2OPO_3^{2-}$
dihydroxyacetone phosphate		glyceraldehyde 3-phosphate

No matter what the starting conditions, the ratio of glyceral dehyde 3-phosphate to dihydroxyacetone phosphate is alwaysfound to be 0.0475 once equilibrium has been attained. By substituting this value of K'_{eq} into Equation 2.4, we obtain

$$\Delta G° = -2.303RT \log_{10} (0.0475) \qquad (6.5)$$

Substituting appropriate values for the gas constant R (0.00198 kcal/mol/K) and absolute temperature T (25°C=298K), we can calculate

$$\Delta G = -2.303 \times 0.00198 \times 298 \times \log_{10} (0.0475)$$

$$= +1.8 \text{ kcal/mol}$$

Hence for every mole of dihydroxyacetone phosphate converted to glyceraldehyde 3-phosphate *under standard conditions*, 1.8 kcal of energy is consumed.

Now let us observe what would happen to this reaction under an arbitrary set of initial conditions such as might exist in the cell, for example, a dihydroxyacetone phosphate concentration of 10^{-4}M and a glyceraldehyde 3-phosphate concentration of 10^{-6}M. The actual free-energy change ΔG can be calculated for these conditions by substituting into equation 2.1 as follows:

$$\Delta G = \Delta G^\circ + 2.303\,RT \log_{10}\frac{[\text{products}]}{[\text{reactants}]} \quad (6.6)$$

$$= 1.8 \text{ kcal/mol} + 2.303 \text{ RT} \log_{10}\frac{10^{-6}}{10^{-4}} \quad (6.7)$$

$$= 1.8 \text{ kcal/mol} + (2.303 \times 0.00198 \times 298 \times -2)$$

$$= 1.8 \text{ kcal/mol} - 2.7 \text{ kcal/mol}$$

$$= -0.9 \text{ kcal/mol}$$

Thus in spite of the positive value of ΔG° previously calculated for the reaction, ΔG turns out to be negative under these particular conditions. This negative value of ΔG means that under the above specified conditions, the reaction will proceed spontaneously with a release of free energy, even though such would not be the case under standard conditions. Since ΔG is calculated on the basis of the prevailing initial concentrations of reactants and products rather than arbitrary standard conditions, ΔG rather than ΔG° determines the direction in which reactions proceed in the cell.

KINETICS OF CHEMICAL REACTIONS

The thermodynamic approach is useful in defining the directions in which reactions tend to proceed, but it tells us nothing about the rates at which they occur. Analysis of the factors involved in determining how fast a reaction will occur is referred to as the study of *kinetics*. Reaction kinetics are ultimately based on the fact that the conversion of reactants to products always involves an intermediate stage, called the *transition state,* whose energy level is higher than that of either the reactants or the products. The term free energy of activation, or ΔG, is used to refere to the difference between the free-energy level of the reactants and the free-energy level of the transition state. The rate at which a chemical reaction proceeds is directly proportional to the number of molecules attaining the high energy transition state. As a result, any factor that increases the number of molecules having sufficient energy to attain the transition state will increase the rate of the reaction.

There are two general ways in which this can be accomplished. The first involves temperature. At moderate temperature a typical reaction mixture consists of a population of molecules with varying energy levels. Although the average free-energy level of the mixture

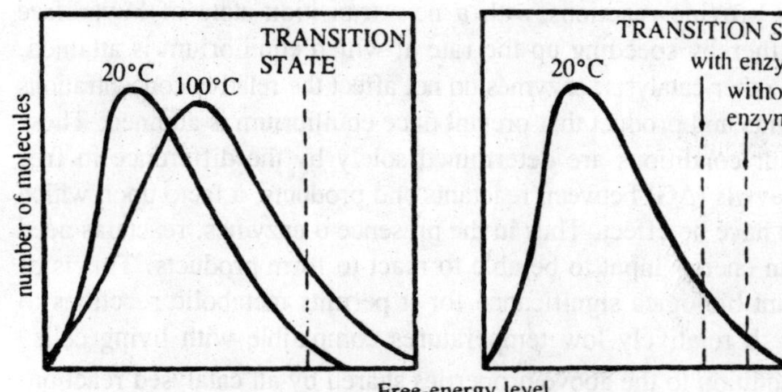

Figure 6.11: Energy distribution diagrams illustrating two ways in which reaction rates can be increased.

is lower than that of the transition state, a small fraction of the molecules has sufficient energy to achieve this activated state. The relative number of molecules in this latter category *determines* how fast the reaction will proceed. Raising the temperature increases the overall thermal energy of the population as a whole, thereby increasing the number of molecules with free-energy levels high enough to enter the transition state. This higher proportion of molecules attaining the transition state in turn increases the rate of product formation.

The second general method for increasing reaction rates involves the use of catalysts. A catalyst combines with one or more reactants to generate a new transition state whose $\Delta G\dagger$ is lower than the $\Delta G\dagger$ of the transition state for the uncatalysed reaction. In other words, the catalyst provides an alternative reaction pathway with a transition state of lower free energy. Hence at any given temperature, the number of molecules possessing sufficient free energy to attain the transition state will be higher for the catalysed reaction than for the uncatalysed reaction.

Because the new transition state created by a catalyst is part of both the forward and reverse reactions, catalysts increase reaction rates in both directions rather than selectively enhancing the rates of either the forward or backward reaction. A catalyst therefore can only speed up the rate at which equilibrium is achieved. The equilibrium concentration of reactants and products are solely determined by the thermodynamic calculation of ΔG, which is not influenced by the presence or absence of a catalyst.

Enzyme-catalysed reactions obey all the basic principles of thermodynamics and kinetics described above. Like other catalysts, enzymes provide reactions with a new transition state of lower free energy, thereby speeding up the rate at which equilibrium is attained. And like other catalysts, enzymes do not affect the relative concentrations of reactants and product that prevail once equilibrium is attained. These equilibriur conditions are determined solely by the difference in free energy levwls (ΔG) between reactants and products, a facto upon which enzymes have no effect. Thus in the presence o enzyines, reactants need less of an energy input to be able to react to form products. This is of paramount biologica significance for it permits metabolic reactions to occur at th relatively low temperatures compatible with living cells.

In addition to the above properties shared by all catalysed reaction, enzymatic reactions exhibit several unusual chara teristics not commonly encountered elsewhere. In the following sections we shall elaborate upon some of these chara teristics and discuss their implications for the mechanism enzyme action.

The Michaelis-Menten Concept of Enzyme Action

What is the relationship between v and [S] when the rate of conversion of S to P is increased by adding a catalyst-an enzyme ? The plot of v versus [S] now takes the form. In contrast to the nonenzymetic reactions we now find three zones in the relationship between v and [S]: (1) at low [S], v is virtually proportional to [S] as in any first-order reaction, except that the rate may be magnitudes greater than observed in either the unheated or heated nonenzymatic reactions ; [2] beyond a certain [S], v increases less, is no longer linearly related to [S], and instead shows the characteristics, of a mixed-order reaction; and (3) finally a value of [S] is reached at which *v* becomes constant, being essentially independent of [S].

It is typical of enzyme reactions that the reaction rate becomes independent of [S] when [S] is increased to a sufficiently high level. This observation, from which it can be deduced that the enzyme can be saturated with substrate, led to the proposal by Michaelis and Menten that the initial and essential step in an enzyme-catalysed reaction is the reversible formation of an enzyme-substrate complex. Accounting for the kinetics of formation and breakdown of such a complex, one can derive a useful mathematical relationship between the reaction observed at any particular [S] and the maximal rate which is achieved at a high [S]. The *Michaelis-Menten equation,* which is developed below, describes the kinetics of any enzyme reaction in which there is only one substrate.

The Michaelis-Menten theory assumes the following sequence of events:

$$E + S \underset{k_2}{\overset{k_1}{\rightleftharpoons}} E.S \underset{k_4}{\overset{k_3}{\rightleftharpoons}} E + P \tag{6.8}$$

It shall also be assumed that k_4 is very small or, in other words, that E . S → P + E is a one-way reaction.

Let us examine first the change in the concentration of E.S. the enzyme-substrate complex, with time. The rate of formation of E.S will be

$$\frac{d(ES)}{dt} = k_1[E_{free}][S] \tag{6.9}$$

Since

$$[E_{total}] = [[E_{free} + ES] \text{ or } [E_{free}] = [E_{total} - ES] \tag{6.10}$$

$$\frac{d[ES]}{dt} = k_1[_{total} - ES][S] \tag{6.11}$$

The rate of breakdown of ES will be

$$\frac{-d[ES]}{dt} = k_2[ES] + k_3[ES] \tag{6.12}$$

When the rate of formation of ES equals its rate od disappearance, the reation system has achieved the steady-state condition. [ES] now remains constant and equations 6.11 and 6.12 can be equated:

$$k_1[E_{total}] - [ES][S] = k_2[ES] + k_2[ES] \tag{6.13}$$

Solving for [ES], we obtain

$$k_1[E_{total}][S] - k_1[ES][S] = k_2[ES] + k_3[ES]$$

$$k_2[E_{total}][S] = k_1[ES][S] = k_2[ES] + k_3[ES]$$

$$= ES[k_1[S] + k_2 + k_3]$$

and $$[ES] = \frac{k_1[E_{total}][S]}{k_1[S] + k_2 + k_3} = \frac{[E_{total}][S]}{[S] + (k_2 + k_3)/k_1}$$

$(k_2 + k_3)/k_1$ will be defined as *KM*. Therefore,

$$[ES] = \frac{[E_{total}][S]}{[S] + k_M} \tag{6.14}$$

[ES] is not a useful term because very often it cannot be determined directly. It will be noted from equation ε-8, however, that the breakdown of ES to product and E is a first-order reaction if k_4 is neglected. The rate of this reaction is termed the *initial velocity* v_o of the enzyme reaction and is equal to K_3[ES].

If both sides of equation (6.14) are multiplied by k_3, the resulting equation will be

$$k_3[ES] = \frac{k_3[E_{total}][S]}{[S]+K_M} = v_0 \qquad (6.15)$$

The maximal initial velocity of the reaction will be achieved when the total E in the system is present as ES. Therefore, it follows that $k_3[E_{total}]$ must equal the maximal initial velocity V_{max}. Hence,

$$v_0 = \frac{V_{max}[S]}{[S]+K_M} \qquad (6.16)$$

This is the Michaelis-Menten equation which describes the plot of v_0 versus [S].

To illustrate the utility of this equation, consider first the meaning of the constant K_M. Assume that the initial reaction rate v_0 is equal to one-half the value of V_{max}. Under this condition

$$\frac{V_{max}}{2} = \frac{V_{max}[S]}{[S]+K_M} \text{ or } \frac{1}{2} = \frac{[S]}{[S]+K_M}$$

$$2[S] = [S]+K_M \text{ or } K_M = [S] \qquad (6.17)$$

Therefore, operationally defined, K_M is equivalent to the [S] at which the initial velocity of the enzyme reaction is half of the V_{max}. As defined in equation 6.16, K_M is a composite dissociation constant for the ES complex, accounting for both its reversible formation and dissociation as well as for the reaction in which ES breaks down to form P and regenerate E. Whenever it can be established for any enzyme reaction that k_3 is negligible compared with k_2, the equation for K_M becomes

$$K_M = \frac{k_2}{k_1}$$

and K_M in this approximation is equivalent to the dissociation constant for the ES complex and is designated as K_s:

$$\frac{[E][S]}{[ES]} = K_s \qquad (6.18)$$

When properly defined, K_M and K_s are useful constants for describing enzyme activities both in tissues and purified preparations. For example, a determination of K_s values permits comparison of affinities of a substrate for an enzyme isolated from different sources or a comparison of affinities of structurally related substrates with a given enzyme.

knowing the value of K_M is useful when one needs to determine how much S to employ to assure maximal velocity. Obviously, when (S) is very small compared with K_M, v_o will vary directly with [S] as a simple first-order function

$$v_0 = \frac{V_{max}}{K_M}[S]$$

On the other hand, if [S] is many times larger than K_M,

$$v_0 = V_{max}$$

When [S] is saturating and all of the E is present as ES, $v_0 = k_3$ [ES] =Vmax. The plot of v_0 versus [E] becomes V_{max} versus [E]:

These are the experimental conditions required for measuring the enzyme level in a particular preparation or for comparing the activities of an enzyme from different sources.

Alternative Algebraic forms of the Michaelis-Menten Equation

The rectangular hyperbolic plot that relates the initial rate of a one-substrate enzyme reaction with the initial substrate concentration is not particularly convenient for determining V_{max} accurately. Two other algebraic forms of the equation are now employed widely in kinetic analyses of enzymatic reactions. One of these transformations is the equation obtained by taking the reciprocal of each side of the Michaelis-Menten equation

$$\frac{1}{v_0} = \frac{K_M + [S]}{V_{max}[S]}$$

or

$$\frac{1}{v_0} = \frac{K_M}{V_{max}[S]} + \frac{[S]}{V_{max}[S]}$$

or

$$\frac{1}{v_0} = \frac{K_M}{V_{max}}\frac{1}{[S]} + \frac{1}{V_{max}} \qquad (6.19)$$

This form of the Michaelis-Menten relationship, referred to as the *Lineweaver-Burk Equation*, is the same equation that describes the straight-line plot of y versus x:

$$y = mx + b$$

in which *m* is the slope of the line and *b* is the intercept on the *y* axis. Hence, plotting $1/v_0$ versus 1/[S], the graphical representative of the Lineweaver-Burk equation.

The stop of the line is K_M/V_{max} and the intercept is $1/V_{max}$. The intercept of the extrapolation of this plot to the $1/[S]$ axis will have the value of $-1/K_M$ because

$$\frac{1}{v_0} = 0 = \frac{K_M}{V_{max}}\frac{1}{[S]} + \frac{1}{V_{max}}$$

and $$-K_M \frac{1}{[S]} = 1$$

and $$\frac{1}{[S]} = -\frac{1}{K_M} \qquad (6.20)$$

It is apparent that the linear double-reciprocal plot of the Lineweaver-Burk equation for determining *Vmax* circumvents the problems of estimating this velocity as a limiting value in the curvilinear plot of v_0 versus [S].

Another commonly employed permutation of the MichaelisMenten equation is that resulting when both sides of the Line weaver-Burk expression are multiplied by v_0

$$\frac{1}{v_0} = \frac{K_M}{V_{max}}\frac{1}{[S]} + \frac{1}{V_{max}}$$

$$\frac{v_0}{v_0} = \frac{K_M v_0}{V_{max}[S]} + \frac{v_0}{V_{max}}$$

$$1 = \frac{R_M v_0}{V_{max}[S]} + \frac{v_0}{V_{max}}$$

$$\frac{v_0}{V_{max}} = 1 - \frac{K_M v_0}{V_{max}[S]}$$

$$v_0 = V_{max} - \frac{K_M v_0}{[S]} \qquad (6.21)$$

The straight-line function described by this equation.

This intercept on the v_o axis is V_{max} and that on the $v_0/[S]$ axisis V_{max}/K_M. The slope of the line is $-K_M$.

Mechanism of Enzyme Actions

There is a very close structural relationship between the molecular surface of an enzyme and its substrates. Enzymes are protein molecules with definite surface geometry. The functional groups of the enzyme are exactly complementary to those of the substrate. Only particular types of substrate molecules will fit with a given enzyme molecule. For example, substrates A and B will fit into the enzyme E but not substrate C. This is referred to as a *'lock-and-key'* mechanism. Thus reactions involving A and B will be speeded up but, not reactions involving C.

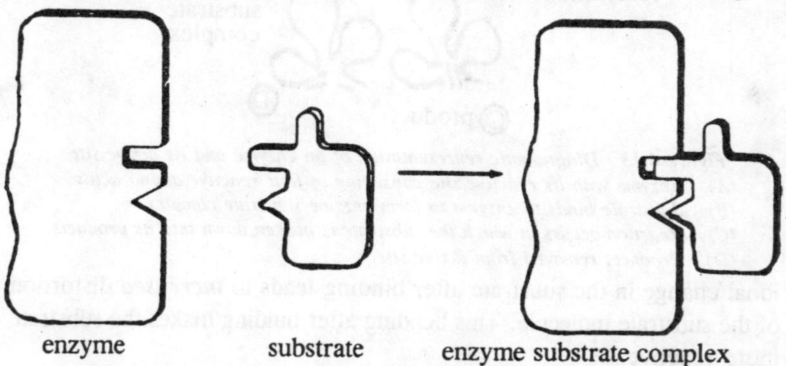

Figure 6.12 : Diagram illustrating how only particular substrates react with an enzyme.

The *enzyme* enters into a chemical combination with the *substrate* to form an *enzyme-substrate complex* (Michaelis-Menten hypothesis).

$$E + S \rightarrow ES$$

The enzyme substrate complex then breaks down to give the products of reaction. The enzyme is released and can be used over and over again.

$$ES \rightarrow E + \text{Products}$$

Active Site

An enzyme has a distinct cavity or cleft in which the substrate is bound. The cleft contains an *active centre* in which the amino acids are grouped together in such a way as to enable them to combine with the substrate. The reactive amino acids may lie widely separated in the polypeptide chain. The chain, however, undergoes folding in such a manner that the reactive amino acids come together in the active site.

It is believed that when the substrate molecule binds to the active site, its parts are hold together in such a way as to cause distortion of the chemical bonds, i.e., the bonds are weakened. This distortion of the chemical bonds of the substrate increases its reactivity, and thus speeds up the rate of the reaction. The products of the reaction are released because they are less firmly bound. The mechanism suggested above has been called the *strain model* of enzyme catalysis.

Another model, called the *rack model,* suppose that the conformat-

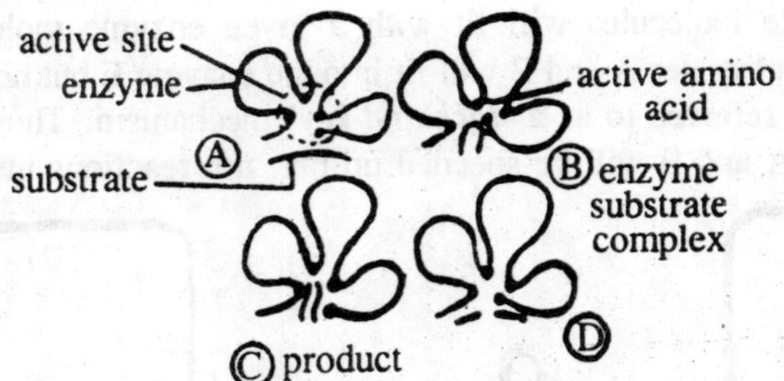

Figure 6.13 : Diagramatic representation of an enzyme and its active site.
(A) Enzyme with its reactive site consisting of four reactive amino acids.
(B) Substrate binds to enzyme to form enzyme-substrate complex.
(C) Reaction occurs in which the substrate is broken down into its products.
(D) Products removed from active site.

ional change in the substrate after binding leads to *increased* distortion of the substrate molecule. This bending after binding makes the substrate more reactive.

Another way in which the reactive site is believed to speed up a reaction is by *excluding water molecules* of the solvent from the site of reaction. The tight fit of the substrate in the active site of the enzyme molecule does not permit water molecules at the site of reaction. It is known that in the case of certain compounds exclusion of water molecules greatly affects the rate of reaction.

Induced Fit Model

This model was given by Koshland. In the Fischer model, i.e., lock-and-key mechanism, the catalytic site is presumed to be preshaped to fit the substrate. In the induced fit theory, the substrate induces a conformational change in the enzyme. This aligns amino acid residues or the other groups on the enzyme in the correct spatial orientation for substrate binding, catalysis or both. At the same time, the other amino acid residues may get buried in the interior of the enzyme. This is depicted in Figure elsewhere in this chapter.

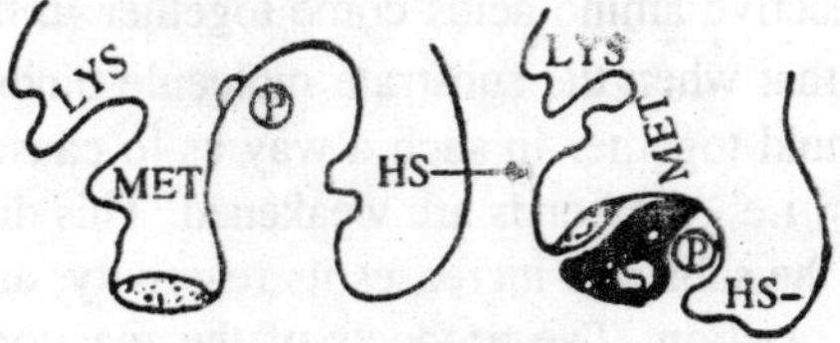

Figure 6.14: Representation of an induced fit by a conformational change in the protein structure.

In the absence of substrate, the catalytic and the substratebinding groups are several bond distances removed from one another. When the substrate approaches, there occurs a conformational change in the enzyme protein, aligning the groups correctly for substrate binding and for catalysis. At the same time there also occurs a change in the spatial orientations of the Figure elsewhere in this chapter. Representation of alternative reaction paths for a substrate induced conformational change.

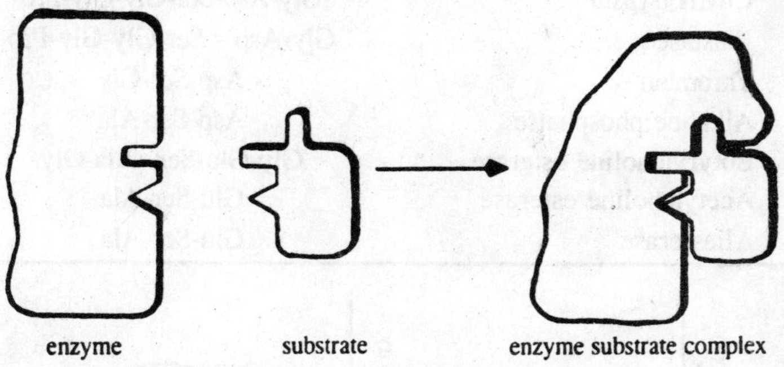

Figure 6.15: Representation of alternative reaction paths for a substrate induced conformational change.

The main evidence in favour of induced fit model comes from demonstration of conformational changes daring substrate binding and catalysis with creative kinase, phosphoglucomutase, and several other enzymes. Upon this time, it could not be established about the exact sequence of events in a substrate-induced conformational change. There may be several possibilities which are depicted.

Even if one knows the complete primary structure of an enzyme, it is not very easy to decide exactly which residues exactly constitute the catalytic site.

FACTORS AFFECTING ENZYME ACTIVITY

Substrate Concentration

Enzyme activity has been found to first increase with increase in concentration of the substrate to a maximum and then it levels off. At this concentration of substrate, amount of enzyme may probably become limiting.

Two types of graphs are obtained from the plotting of substrate concentration and enzyme activity. Straight line or Michaelis type of curve is obtained with the enzymes if they are made up of one unit.

If the enzymes are made up of the several sub units and they exhibit interaction amongst subunits (allosteric enzymes) then igmoidal type of curve is obtained.

Table 6.2: Amino acid sequence of active sites of various enzymes.

Enzyme	*Sequence of Active Site*
Trypsin	Gly-Asp-Ser-Gly-Gly-Pro
Chymotrypsin	Gly-Asp-Ser-Gly-Gly-Pro
Elastase	Gly-Asp - Ser-Gly-Gly-Pro
Thrombin	Asp-Ser-Gly
Alkaline phosphatse	Asp-Ser-Ala
Butyrylcholine esterase	Gly-Glu-Ser- Ala-Gly
Acetylcholine esterase	Glu-Ser-Ala
Aliesterase	Glu-Ser-Ala

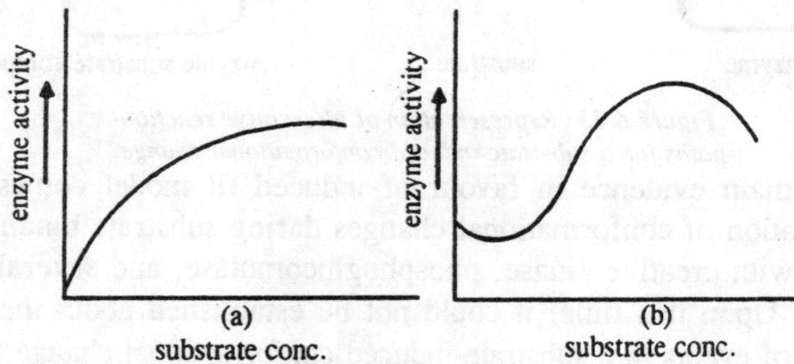

Figure 6.16 : Effect of substrate concentration on enzyme activity, a—Michaelis-Menten type curve. b—Sigmoidal curve.

Concentration of Enzymes

Increase in enzyme concentration has been found to increase enzyme activity. When an excess of substrate is present, increasing the enzyme concentration by two times generally doubles the rate of formation of the product. With the given concentration of the enzyme also, a point may be reached where all the substrate molecules get bound to the enzyme and further increase in enzyme does not have any effect on the. formation of the product.

Effect of pH

As enzymes are made up of proteins, they are vety sensitive to change in pH. A graph of enzyme activity against pH is generally bell shaped. There is a optimum pH, which depends upon the nature of the enzyme. However, some enzymes may be having two optima of pH or no sharp pH but a range of pH.

At optimum pH the activity of enzymes is maximum. For most enzymes the effective pH range is 4-9, Beyond these limits denaturation

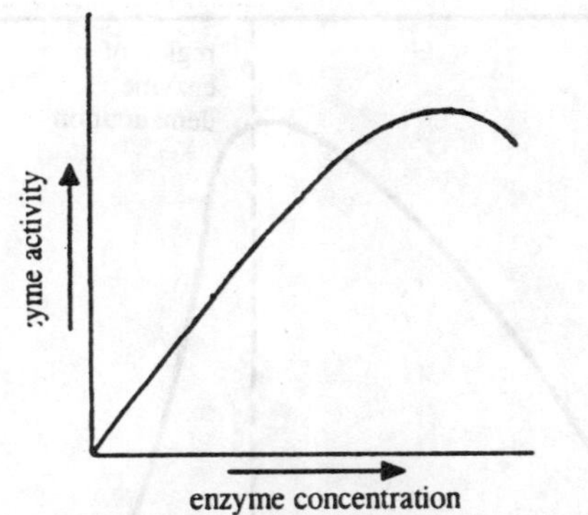

Figure 6.17: Effect of enzyme concentration on enzyme activity.

of enzymes takes place. The optimum pH for pepsin is 2.0 and for trypsin 8.0.

Effect of ions

Many enzymes only become active in the presence of a cation such as Mg^{2+}, Ca^{2+}, Mn^{2+}, Zn^{2+}, Na^+ or K^+. In some cases, the cations get loosely bound to the enzyme, while in other cases they get bound to the substrate. Anions have also been found to increase the enzyme activity. For example, chloride enhances the activity of salivary amylase. Concentration of ions also influences the enzyme activity.

Effect of Temperature

Enzyme action is greatly affected by temperature. If the temperature is increased by 10°C the rate of most chemical reactions is doubled.

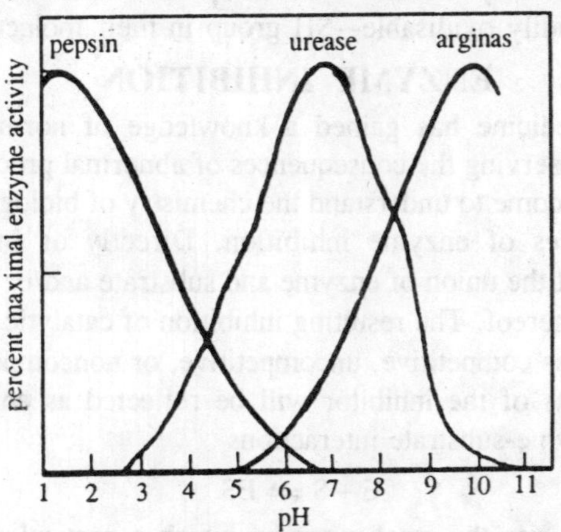

Figure 6.18: Effect of varying pH on the catalytic activity of three different enzymes.

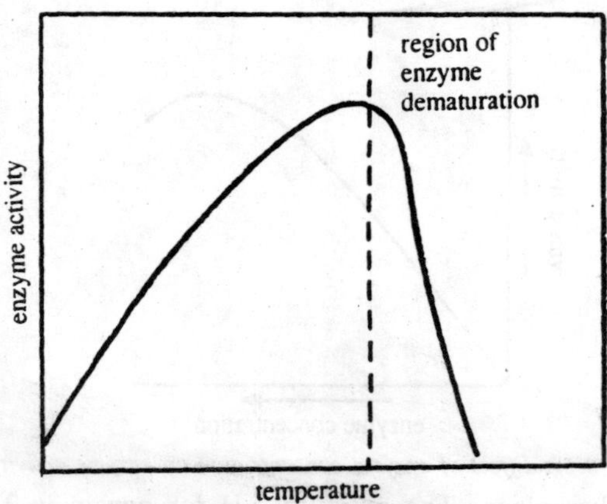

Figure 6.19 : Effect of increasing temperature on the rate of an enzyme-catalysed reaction.

However, at 40°C-60°C there is loss of enzyme activity because denaturation of proteins occurs at this temperature. There are few enzymeş which continue to be active above 60°C also. The enzymes of dry tissues such as seeds and spores, are more resistant to high temperature.

Redox Potential

Enzymes have been found to be sensitive to the redox potential of the cell also. Oxidising and reducing enzymes also take part in changing the redox component of a cell and hence they get affected by the redox potential. Some enzymes are affected by redox potential due to the presence of readily oxidisable—SH group in their molecule.

ENZYME INHIBITION

Just as medicine has gained a knowledge of normal metabolic pathways by observing the consequences of abnormal processes, so the biochemist has come to understand the chemistry of biological catalysis through analyses of enzyme inhibition. Directly or indirectly, all inhibitors affect the union of enzyme and substrate and/or the catalytic consequences thereof. The resulting inhibition of catalytic function can be categorised as competitive, uncompetitive, or noncompetitive. In all cases the effects of the inhibitor will be reflected as changes in the kinetics of enzyme-substrate interactions

$$E + S \rightleftharpoons ES$$

To characterise the mechanism by which a particular inhibitor is adversely affecting an enzyme reaction, questions such as the following must be answered. Is the degree of inhibition at a fixed enzyme concentration and fixed inhibitor concentration dependent on [S]? Is the degree of inhibition time dependent, suggesting that the inhibitor may be combining irreversibly with the enzyme?

Consider first the simplest type of inhibition—that due to a reversible competition between the inhibitor and substrate for binding at the substrate-binding site.

SUBSTRATE-COMPETITIVE INHIBITION

When a competitive inhibitor binds reversibly witli the enzyme, the complex formation can be described in the same manner as the anzyme-substrate interaction

$$E + I \rightleftharpoons EI$$

The dissociation constant of the complex, K_i, can be defined as

$$K_i = \frac{[E][I]}{[EI]} \quad (6.22)$$

Since the inhibitor is not chemically altered, the only fate of the El complex is to dissociate into E and I. K_i is, therefore, a true dissociation constant and is comparable to K_i. The simplest experimental test of substrate-competitive inhibition is to determine whCther the degree of inhibition at a fixed inhibitor concentration can be decreased by increasing the substrate concentration. Referring again to the relationship between v_0 and [S], the presence of a fixed amount of inhibitor with varying amounts of S will give the following plots:

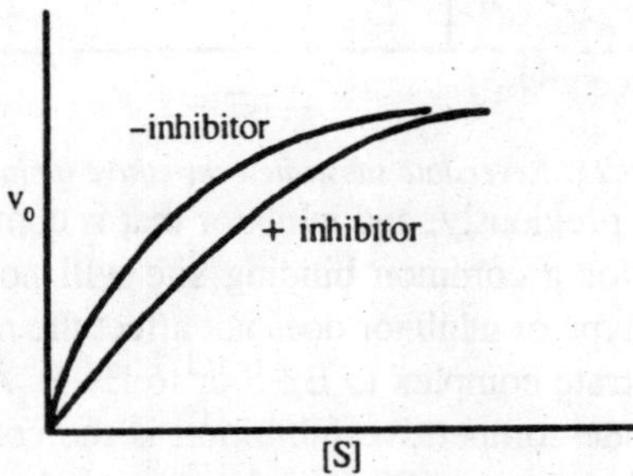

As would be predicted from the Michaelis-Menten model, a higher [S] is required to reach half-maximal velocity for the inhibited reaction than for the uninhibited reaction. Theoretically, the inhibition can be completely reversed by making the [S] [1] ratio very high ; the two curves will merge asymptotically as V_{max} is approached.

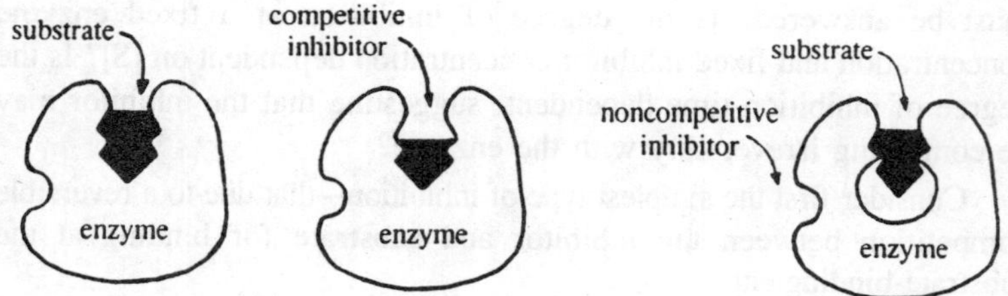

Figure 6.20: Distrinction between a competitive inhibitor and a non-competitive inhibitor.

Just as the kinetic parameters of the Michaelis-Menten relationship are conveniently evaluated by a double-reciprocal plot (equation 6.22), the kinetics of an inhibited enzyme reaction can also be analysed by plots oi 1/[S] versus $1/v_0$. Thus, when the [I] is kept constant and the [S] is varied, the uninhibited and co npttitively inhibited reactions will be related as shown in Figure. The extrapolated intercepts will be

$$(a) -\frac{1}{K_M} \text{ and } (b) -\frac{1}{K_M(1+[I]/K_M)}$$

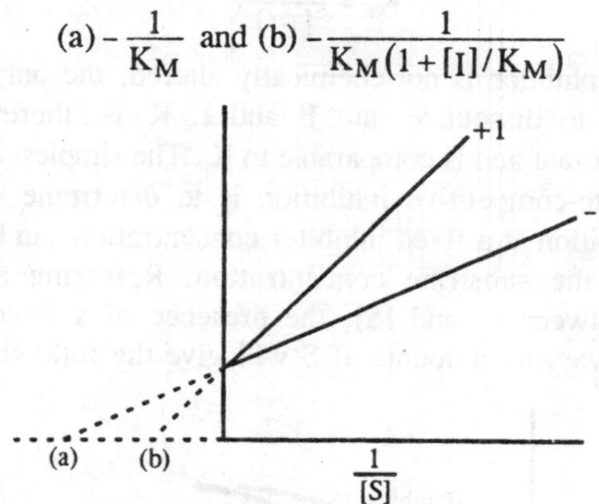

Figure 6.21: Reversible substrate-conpetitive inhibition.

As emphasized previously, and inhibitor that is competing reversibly with the substrate for a common binding site will not affect the V_{max} of the system. This type of inhibitor does not affect the rate of breakdown of the enzyme-substrate complex to E+S or to E+P. Accordingly, one criterion for substrate-competitive inhibition is the common intercept. $1/V_{max}$, in plots of $1/v_0$ versus 1/[S]. Corroborative evidence for concluding that an inhibition is of the single substrate-competitive type and also reversible is obtained by employing a range of fixed levels of inhibitor in kinetics analyses such as those shown in Figure elsewhere in this chapter.

The mathematical formulation describing the reversible competitive between inhibitor and substrate must account for the following processes:

$$E + S \rightleftharpoons ES \rightleftharpoons E + P$$
$$E + I \rightleftharpoons EI \text{ (inactive)}$$

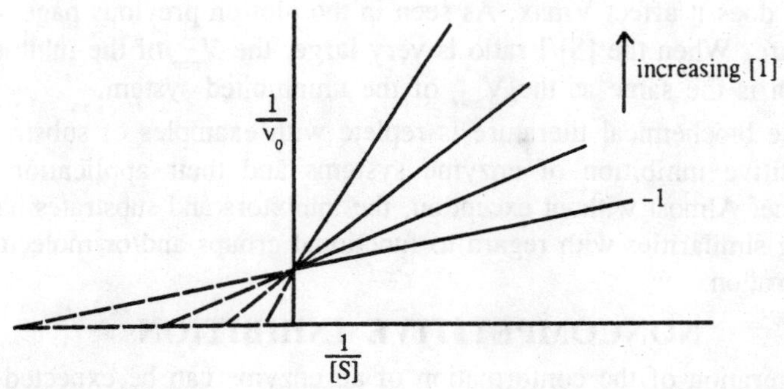

The final form of the Michaelis-Menten relationship in this case is

$$\frac{1}{v_0} = \frac{K_M}{V_{max}}\left(1 + \frac{[I]}{K_i}\right)\frac{1}{[S]} + \frac{1}{V_{max}} \qquad (6.23)$$

In the reciprocal plot of vo versus [S] of the inhibited reaction, the intercept on the ordinate, I/V_{max}, remains the same as for the uninhibited reaction as pointed out above. The slope of the line now becomes

$$\frac{K_M}{V_{max}}\left(1 + \frac{[I]}{K_I}\right)$$

Compared to the uninhibited reaction the slope has been increased by a factor of $I+[I]/K_i$. Equation 6.23 can also be used to calculate K_i, the dissociation constant of the inhibitorenzyme complex. As will be discussed in a later section, a knowledge of the values of K_m and K_i is not only relevant to our understanding of enzyme mechanisms but also has application in the design and clinical use of chemotherapeutic agents.

It was noted in comparing the hyperbolic plots of v_0 versus [S] in the substrate-competitive inhibitor system that to attain the same v_0 in both the uninhibited and inhibited reactions, the [S] in the inhibited reaction must be higher than in the uninhibited system. This is a reasonable consequence of the two competiting reactions

$$E + S \rightleftharpoons ES \rightleftharpoons E + P$$

and

$$E + I \rightleftharpoons EI$$

At any given moment some of the enzyme, being in the form of EI, is unavailable for ES formation. It is important to realise, however, that the presence of I affects neither the association of E and S nor the dissociation of the ES ooniplex. The value of K_M is the same in the presence of I as in its absence. Just as the I does not alter the Km, neither does it affect Vmax. As seen in the plot on previous page and in Figure. When the [S]/I ratio is very large, the V_{max} of the inhibited reaction is the same as the V_{max} of the uninhibited system.

The biochemical literature is replete with examples of substrate-competitive inhibition of enzyme systems and their application to medicine. Almost without exception, the inhibitors and substrates have striking similarities with regard to functional groups and/or molecular conformation.

NONCOMPETITIVE INHIBITION

Alteration of the conformation of an enzyme can be expected to affect its catalytic activity. Any reagent that can combine reversibly with the enzyme in some region or with some functional group outside the substrate-binding site may cause a structural change in the protein if the affected region or site is needed to maintain the native shape of the molecule. Such a binding site. The kinetic consequences of such structural changes can be both a decreased rate of ES formation as well as a decreased rate of ES breakdown to E and reaction product(s). The inhibitor can combine with either E or ES:

$$E + I \rightleftharpoons EI$$

and

$$ES + I \rightleftharpoons ESI$$

The dissociation constants of these inhibitor complexes are

$$K_{EI} = \frac{[E][I]}{[EI]} \tag{6.24}$$

and

$$K_{ESI} = \frac{[ES][I]}{[ESI]} \tag{6.25}$$

It may also be assumed that El can react with S

$$EI + S \rightleftharpoons ESI$$

The K for this reaction, K'_{ESI}, may not have the same value as the K_{ESI} for reaction 6.25. Thus, the "*composite kinetics*" of a noncompetitively inhibited enzyme reaction is complicated. Assuming that the three K values cited above are identical and that El and ESI are inactive, the reciprocal form of the Michaelis-Menten relationship is

$$\frac{1}{v_0} = \frac{K_M}{V_{max}}\left(1+\frac{[I]}{K_I}\right)\frac{1}{[S]} + \frac{1}{V_{max}}\left(1+\frac{[I]}{K_I}\right) \qquad (6.26)$$

The plot of this equation is shown in Figure. The common intercept obtained by extrapolation of the inhibited and noninhibited plots is -1/ K_M. Although the K_s for the two reactions

$$E + S \rightleftharpoons ES \quad \text{and} \quad EI + S \rightleftharpoons ESI$$

can be expected to be different because the active site has been deformed by the noncompetitive inhibitor, the K_M for the two systems can be the same, Why? Assuming that both the El and ESI complexes are inactive, it is as if the noncompetitive inhibitor has merely removed some of the enzyme. That is, the v_0 and V_{max} for the inhibited enzyme are the velocities that would be obtained if the kinetic analysis of the uninhibited control had been repeated using less enzyme.

One test for reversible, noncompetitive inhibition is that the effects of the inhibitor cannot be reversed no matter how much the [S] is increased. The value of V_{max} cannot be increased to that of the uninhibited reaction. The inhibition can only be reversed by removing the inhibitor.

UNCOMPETITIVE INHIBITION

Uncompetitive is the term used to describe an inhibition of enzyme activity resulting from the reversible combination of the inhibitor with the ES complex

$$ES + I \rightleftharpoons ESI \qquad (6.27)$$

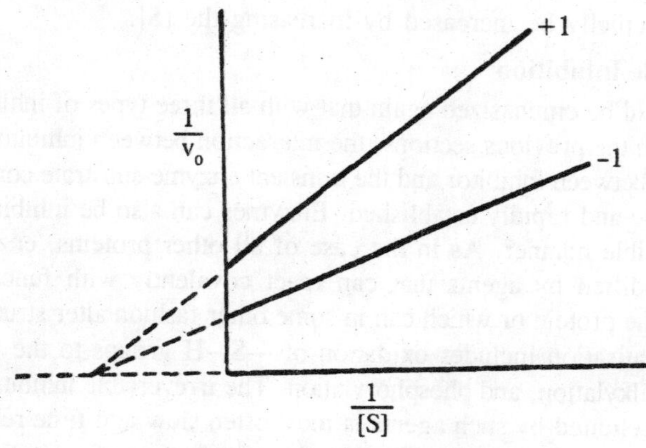

Figure 6.22 : Noncompetitive inhibition.

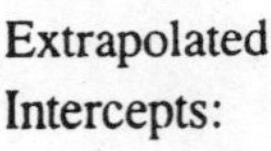

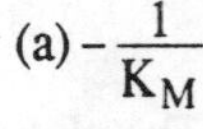

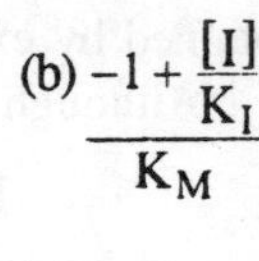

$\frac{1}{v_0}$ +1 -1 (b) (a) $\frac{1}{[S]}$

Figure 6.23 : Uncompetitive inhibition.

The ESI is inactive and the dissociation constant is

$$K_I = \frac{[ES][I]}{[ESI]} \tag{6.28}$$

In the reciprocal form, the relationship between the v_0, [S], and dissociation constants for this type of inhibition is

$$\frac{1}{v_0} = \frac{K_M}{V_{max}} \frac{1}{[S]} + \frac{1}{V_{max}}\left(1 + \frac{[I]}{K_I}\right) \tag{6.29}$$

The plot of initial velocities versus [S] in a system containing a fixed concentration of uncompetitive inhibitor is shown in Figure. Although the V_{max} of the inhibited reaction will be less than for the uninhibited system, the K_M of the substrateenzyme complex itself is not affected by an = rcompetitive inhibitor. From equation 6.29 it will ba seen that the degree of inhibition by a fixed concentration of uncompetitive inhibitor can actually be increased by increasing the [S].

Irreversible Inhibition

It should be emphasized again that with all three types of inhibition discussed in the previous sections, the interaction between inhibitor and enzyme or between inhibitor and the transient enzyme-substrate complex is reversible and rapidly established. Enzymes can also be inhibited in an irreversible manner. As in the case of all other proteins, enzymes can be modified by agents that can react covalently with functional groups in the protein or which can in some other fashion alter structure. Such derivatisation includes oxidation of —S—H groups to the —S—S— form, alkylation, and phosphorylation. The irreversible inhibition of an enzyme elicited by such agents is most often slow and time related, usually increasing with time. The kinetics of this type of inhibition do not fit the Michaelis-Menten model.

ENZYME REACTIONS INVOLVING MORE THAN ONE SUBSTRATE

Most metabolic reactions are two-substrate enzymatic reactions. For example, both the α-amnio acid and the a-keto acid are substrates for the transaminase that catalyses the transfer of the amino group of an amino acid to an α-keto

$$\alpha\text{-keto acid}_A + \alpha\text{-amino acid} \xrightleftharpoons{\text{transaminase}} \alpha\text{-amino acid}_A + \alpha\text{-keto acid}_B$$

Although the overall kinetics of such a system are more complicated than for the process $E+S \rightleftharpoons ES \rightarrow E+P$, the Km for each substrate in a two-substrate system can be determined if the reaction is made pseudo first order. This is accomplished by keeping the concentration of one substrate constant, at its saturation level, and then measuring initial rates of reaction at various concentrations of the second substrate. The analysis is repeated, with the conditions being reversed, to obtain the second K_M.

Having determined the K_M for each of the two substrates, one can then carry out kinetic analyses of the reaction under conditions in which the concentration of one substrate (e.g., amino acid_A) is varied while the concentration of the other (e.g., keto acid_B), is kept constant and lower than its saturating level. The analysis is repeated with several concentrations of the second substrate (i.e., in this case, keto acids). Plots of $1/v_0$ versus $1/[\text{amino acid}_A]$ give a series of parallel lines:

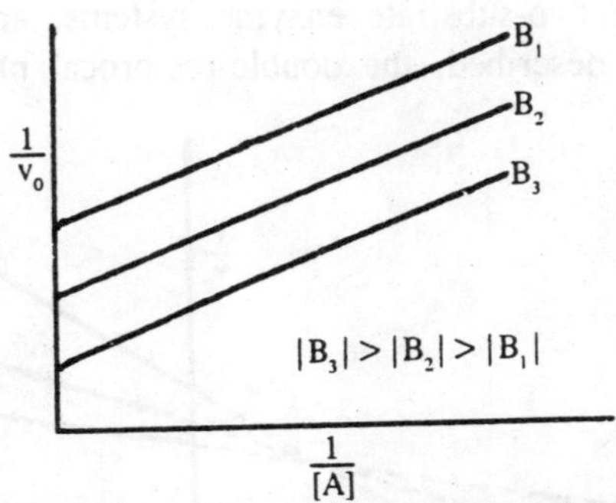

One substrate must be bound to the enzyme and one of the reaction products must be released before the second substrate can enter and the second product can be released. The enzyme, and/or its prosthetic group or coenzyme, is transiently and reversibly modified during the total reaction. The sequence of reactions can be summarised as follows [the transaminase bound to the aldehydic and amino forms of its coenzyme are designated as $E-\overset{H}{C}=O$ and $E-\underset{H}{\overset{H}{C}}-NH_2$]:

$$\text{amino acid}_A-NH_2+E-\overset{O}{\overset{||}{C}}H \rightleftharpoons \text{amino acid}_A-NH_2.E-\overset{O}{\overset{||}{C}}H$$

$$\text{amino acid}_A-NH_2.E-\overset{O}{\overset{||}{C}}H \rightleftharpoons \text{keto acid}_A+E-\underset{H}{\overset{H}{C}}-NH_2$$

$$E-\underset{H}{\overset{H}{C}}-NH_2+\alpha\text{-keto acid}_B \rightleftharpoons E-\underset{H}{\overset{H}{C}}-NH_2.\alpha\text{-keto acid}_B$$

$$E-\underset{H}{\overset{H}{C}}-NH_2.\alpha\text{-keto acid}_B \rightleftharpoons E-\overset{O}{\overset{||}{C}}H+\text{amino acid}_B-NH_2$$

Note that in this mechanism only binary complexes are involved.

The double-reciprocal form of the rate equation that describes each of the straight-line plots shown above is

$$\frac{1}{v_0}=\frac{K_M^A}{V_{max}}\frac{1}{[A]}+\left(1+\frac{K_M^B}{[B]}\right)\frac{1}{V_{max}} \qquad (6.30)$$

With some two-substrate enzyme systems, analysed under the conditions just described, the double-reciprocal plots may take the following form

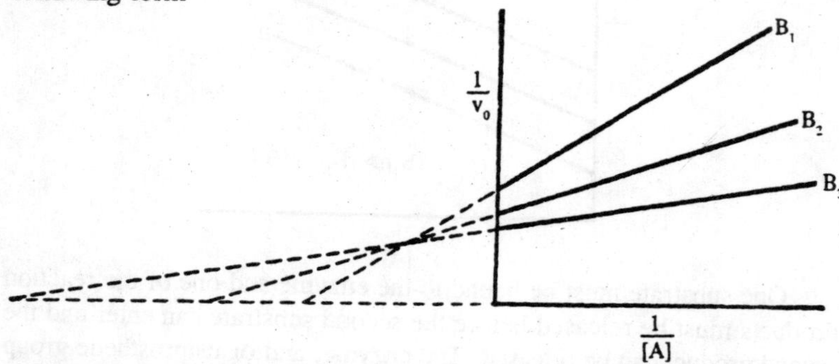

These results permit the interpretation that the reaction can proceed only if both substrates have been bound to the enzyme's active site, to

forma ternary complex. A common example of an enzyme reaction which has been demonstrated to utilise this type of mechanism is the NAD^+-dependent dehydrogenation of malate to oxaloacetate, one of the reactions of the Krebs cycle. In this system, the combination of NAD^+ and malate with the dehydrogenase protein does not occur randomly; the NAD^+ is bound prior to the malate:

$$NAD^+ + \text{dehydrogenase} \rightarrow \overset{+}{NAD}\text{-dehydrogenase}$$

$$\overset{+}{NAD}\text{-dehydrogenase} + \text{malate} \rightarrow \overset{+}{NAD}\text{-dehydrogenase-malate}$$

Following the transfer of electrons from the malate to the NAD^+ within the ternary complex, the release of products is similarly ordered; the NADH is the last reaction product to leave the complex. This sequence of substrate binding and product release is consistent with the observation that the dehydrogenation of malate is found to be inhibited in the presence of excess NADH. NADH can complete with NAD^+ and thereby hinders the normal rate of formation of the active ternary complex.

In contrast to the mechanism involving an ordered synthesis and breakdown of a ternary complex, some two-substrate enzyme reactions proceed via a complex in which substrate addition to the enzyme, and product discharge from the complex, occur randomly. For example, the transfer of a phosphate group from phosphocreatine to ADP, by the action of creatine kinase (CPK), occurs in such a nonordered manner. The overall reaction,

Phosphocreatine $(CP) + ADP \xrightleftharpoons{CPK}$ creatine (C) + ATP requires formation of a ternary complex: either

$$CP + CPK \rightleftharpoons CP.CPK$$

$$CP.CPK + ADP \rightleftharpoons ACP.CPK.ADP$$

or

$$ADP + CPK \rightleftharpoons ADP.CPK$$

$$ADP.CPK + CP \rightleftharpoons ADP.CPK.CP$$

The transfer to the phosphate group from CP to ADP then occurs within the ternary complex and the products are dischasged randomly: either

$$CP.CPK.ADP \rightarrow C.\underset{\underset{C+CPK}{\downarrow}}{CPK} + ATP \text{ or } \underset{\underset{ATP+CPK}{\downarrow}}{ATP.CPK} + C$$

Whether the enzyme reaction involving a ternary complex is ordered or random, the Micbaelis-Menten relationship that describes the double-reciprocal plots shown will be:

$$\frac{1}{v_0} = \frac{1}{v_{max}}\left[K_M^A \frac{K_S^A \; K_M^B}{[B]}\right]\left[\frac{1}{[A]}\right] + \frac{1}{V_{max}}\left[1 + \frac{K_M^B}{[B]}\right]$$

In this equation, K_S^A is the composite constant for the dissociation of EA and EAB:

$$\text{EA and EAB} \rightleftharpoons \text{E+EB+A}$$

The extrapolated convergenec point of the plots in the figure represents $-1/K_S^A$.

Coenzymes

All enzymes are protein enzymes are simple proteins whereas others are conjugated proteins. The non-protein part of a conjugated protein or enzyme is known as prosthetic group. It is observed that prosthetic group is necessary for protein to act as an enzyme and is generally termed as *coenzyme*, the protein part of a conjugated or protein as *apoenzyme* and the intact molecule or conjugated protein or *holoenzyme. Thus*,

$$\text{Holoenzyme} \rightleftharpoons \text{Apoenzyme + Coenzyme}$$

Thus, the coenzyme may be defined as a non-proteinous substance necessary for the activity of the enzyme.

Only when enzyme and coenzyme are present together, catalysis will occur.

It is often difficult to distinguish between what is meant by the terms "*coenzymes*", "*prosthetic group*" and "*activators*". Perhaps it is best to reserve the "*coenzyme*" for the organic fragments that are not normally isolated with the apoenzyme. A "prosthetic group" would then represent a firmly attached coenzyme such as iron protoporphyrin and "activator" might be reserved for metals, non-specific reducing agents, and so forth. At any rate, all these substances are generally of low molecular weights and are dialysable and heat stable.

Coenzymes are organic molecules of a size intermediate between the small-molecule intermediary metabolites, which serve as the substrates of enzymatic reactions, and the macromolecular proteins. A coenzyme is an easily dissociable portion (sometimes called the prosthetic portion or group) attached to the protein component (apoenzyme) to form the complete, enzymatically active, conjugated protein (holoenzyme). Each coenzyme (or cofactor) acts usually as acceptor or donor of some specific type of atom or group of atoms to be removed from or added to a small molecule substrate in a reaction cataly sed by the hole-enzyme. For example, the folic acid coenzymes accept or donate single carbon units (at various states of oxidation) in a considerable number of enzymatic reactions grouped together as single carbon unit metabolism.

It is often helpful to regard the coenzyme a second substrate i.e., a cosubstrate. This is attributed to two reasons

(i) The chemical changes in the coenzyme exactly counterbalance those taking place in the substrate. An example is in transphosphorylation reactions involved in the metabolism of sugars, for every molecule of sugar phosphorylated, one molecule of ATP is dephosphorylated and converted to ADP.

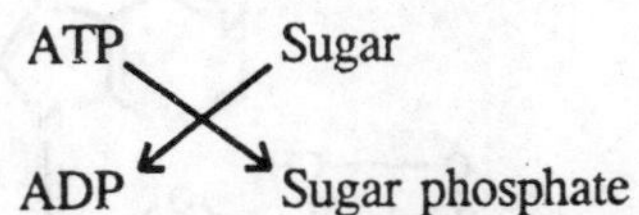

(ii) The second reason for the importance of co-enzymes is that their reactions may be of physiological importance. For example, the importance of ability of muscle to convert anaerobically pyruvate to lactate resides neither in pyruvate nor in lactate themselves but the reaction merely serves to convert NADH to NAD^+.

Without NAD^+, glycolysis cannot occur and anaerobic ATP synthesis ceases and hence muscular work ceases.

Many coenzymes are closely related to *vitamins* and are the derivatives of vitamins. The B group vitamins (except biotin and lipoic acid) function as parts of some coenzyme, e.g., *CoA* is a derivative of *pantothenic acid*. When there is vitamin deficiency the coenzyme concentration decreases. Consequently enzyme function is depressed. Nucleotides may also function as coenzymes in certain metabolic reactions.

In some cases enzymes may be activated by simple substances 'like metal ions which are then called *activators*. The substrate forms a complex with the metal ion and then reacts with the enzyme examples of metal ions which function as cofactors are Na^+, K^+, Ca^{2+}, Co^{2+}, Mg^{2+}, Mn^{2+}, Cd^{2+}, Fe^{2+}, Cr^{3+} and Al^{3+}.

Structure of Coenzymes

Metallic ions, vitamins and other types of organic compounds have been found to function as prosthetic groups or coenzymes. An essential component of most of coenzymes is generally phosphate in the form of nucleotides. For example, nicotinamide adenine dinucleotides,. also known as coenzyme I (NAD) or coenzyme II (NADP) act as coenzymes for dehydrogenases. Iron containing cytochromes make the electrons to get transferred from flavorproteins to cytochrome oxidase. Pyruvate decarboxylase is an enzyme that splits pyruvate into CO_2 and acetaldehyde in the yeast cells. It contains thiamine (vatamin B_1) as a part of its coenzyme.

Classification of Coenzymes

The coenzymes have beenclassified in accordance to the reactions they participate in. Such a classification with their corresponding vitamins has been summarised in Table elsewhere in this chapter.

NH_2
ADENINE
PHOSPHATE HO—P=O
D-RIBOSE
PHOSPHATE HO—P=O
NICOTINAMIDE
$CONH_2$
D-RIBOSE

Coenzyme I (NAD^+)

Mechanism of Action of Coenzymes

Now we will describe the chemical structure *and* mechanism of action of some important coenzymes.

Coenzyme I

It is also known as *coenzymase,* DPN (diphosphopyridine phosphate nucleotide), codehydrogenase I or NAD^+ (nicotinamide adenine dinucleotide).

Coenzyme I is mainly involved in many hydrogen transferring reactions. In these reactions, H is added to a substrate from C-4 atom of the pyridine nucleus. The pyridine gets reduced thereby retaining only two double bonds.

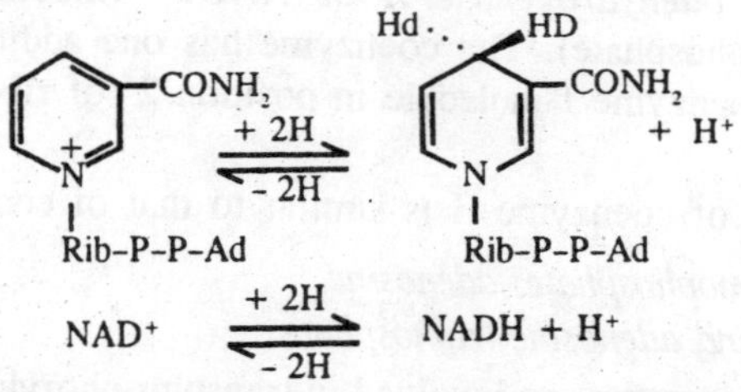

$$\text{Pyridinium-}CONH_2 \text{ (Rib–P–P–Ad)} \underset{-2H}{\overset{+2H}{\rightleftharpoons}} \text{Dihydropyridine-}CONH_2 \text{ (Hd, HD; Rib–P–P–Ad)} + H^+$$

$$NAD^+ \underset{-2H}{\overset{+2H}{\rightleftharpoons}} NADH + H^+$$

From the experimental work, it could be concluded that NAD^+ enzyme complex is *usually stereospecific*, only one hydrogen (H_a or H_b) is reacting exclusively. Which face of NAD^+ is attacked and which hydride ion from NADH is transferred depends upon the nature of the enzyme.

NADH absorbs light at 340 nm whereas NAD^+ shows no absorption at that wavelength. Thus, when NADH is formed during a reaction, absorption of light at 340 nm increase. The increase in absorption has been found to be proportional to the enzyme activity. This property finds use extensively in the bioassay of dehydrogenase and other oxidoreductases which employ NADH as coenzymes.

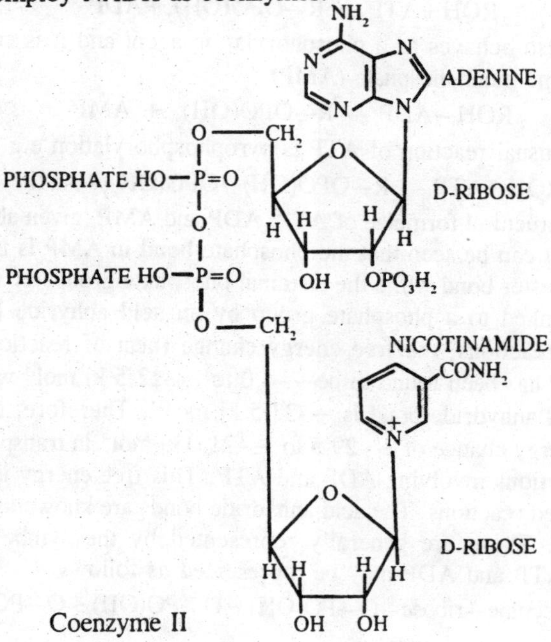

ADENINE

PHOSPHATE HO—P=O

D-RIBOSE

PHOSPHATE HO—P=O

OH OPO_3H_2

NICOTINAMIDE

$CONH_2$

D-RIBOSE

OH OH

Coenzyme II

Coenzyme II

It is also known as *phosphocoenzymase*, TPN (triphosphopyridine nucleotide), *codehydrogenase II* or $NADP^+$ (nicotinamide adenine dinculeotide phosphate). The coenzyme has one additional phosphate group than coenzyme I molecule in position 2' of ribose molecule of adenosine.

The role of coenzyme II is similar to that of coenzyme I.

Adenosine monophosphate, adenosine diphosphate and adenosine triphosphate.

These coenzymes are involved in transphosphorylation.

Adenosine tripbophate is a polyelectrolyte molecule having four negative charges. Since it has low molecular weight, it moves freely in cells. It acts as a mediator to receive energy from one reaction and transfers this energy to drive another reaction. ATP is synthesised in several metabolic reactions of the cell (cyclic, noncyclic phosphorylation and oxidative phosphorylation).

In order to overcome the energy barriers, the energy must be supplied. In biosynthetic processes, ATP supplies this energy in transphosphorylation reactions in the presence of a suitable enzyme and it is converted into adenosine diphosphate (ADP).

$$ROH + ATP \rightarrow R{-}OPO(OH)_2 + ADP$$

ADP also behaves as a phosphorylating agent and it is converted into adenosine monophosphate (AMP).

$$ROH{-}ADP \rightarrow R{-}OPO(OH)_2 + AMP$$

A less usual reaction of ATP is pyrophosphorylation e.g.,

$$ROH + ATP \rightarrow R{-}OPO(OH){-}OPO(OH)_2 + AMP$$

If the structural formulae of ATP, ADP and AMP (given above) are inspected, it can be seen that the phosphate bond in AMP is linked by the normal ester bond while the terminal phosphate groups in ADP and ATP are linked to a phosphate group by an acid anhyride bond. In hydrolytic reactions, the free energy change (heat of reaction) of an easter bond has been found to be ~ -4.0 to ~ -12.5 kj mol^{-1} while that for the acid anhydride bond is ~ -33.5 kj mol^{-1}. Therefore, there is a net free energy change of ~ -29.5 to ~ -21.0 kj mol^{-1} in transphosphorylation reactions involving ADP and ATP. This free energy is used to drive coupled reactions. The acid anhydride bonds are known as energy-rich bonds. These are generally represented by the symbol $\sim$. For example, ATP and ADP may be represented as follows:

ATP : Adenine—ribose—O—PO(OH) $\sim$ O—PO(OH) $\sim$ O—$PO(OH)_2$

ADP: Adenine—ribose—O—PO(OH) ~ O—PO$(OH)_2$

Pyridoxal phosphate (codecarboxylase).

It is involved in a number of important metabolic reactions of the a-amino acids, e.g., transamination, racemisation, decarboxylation and elimination reactions.

O ‖ HO—P—OCH_2 | OH — CHO, OH, N, CH_3

pyridoxal phosphate (PALP)

Role

Nearly all pyridoxal enzymes catalyse reactions of amino acids racemisation, decarboxylation, transamination, elimination of substituents and others. All the reactions can be described by the same machanism.

Pyridoxal peosphate is attached to the apoenzyme tightly; the aldehyde group of pyridoxal reacts with amino acids to form a Schiff's base. For racemisation, there occurs a shifting of electrons in Schiff's base, thereby leading to labilisation of all bonds around the α-carbon atom of the amino acid.

Pyridoxal phosphate also acts as a coenzyme reaction between amino acids and keto acids.

$$\text{R–HCNH}_2\text{–COOH} + \text{R–CO–COOH} \rightleftharpoons \text{R–CO–COOH} + \text{R–CHNH}_2\text{–COOH}$$

This reaction, of great importance biologically is catalysed by pyridoxal phosphate and its mechanism is similar to racemisation.

Thiamine pyrophosphate (cocarboxylase).

It is involved in the transfer of acyl and carboxyl groups. It is derived from the vitamin, thiamine (vitamin B_1).

NH_2 — CH_2—N — CH_3 — N, H_3C, N — $CH_2.CH_2O.$P(=O)(OH)—O—P(=O)(OH)—OH

carboxylase

Action

The enzyme *carboxylase* breaks down pyruvic acid into acetaldehyde in fermentation and carbohydrate metabolism. This reaction takes place in the presence of coenzyme, *cocarboxylase*.

$$\underset{\text{pyruvic acid}}{CH_3COCOOH} \xrightarrow[\text{cocarboxylase}]{\text{carboxylase}} CH_3CHO + CO_2$$

The mechanism of the cocarboxylase depends upon the ionisation of the proton at C—2 in the thiazolium ring. The liability of this hydrogen atom is by its ready displacement by deuterium when thiazolium salts are dissolved in acidified deuterium oxide.

The most important reaction catalysed by TPP is the decarboxylation of pyruvic acid to acetaldehyde.

The decarboxylation of pyruvic acid by this mechanism involves a non-enzymatic process and in it there occurs transfer of electron pairs. In all other thiamine catalysed enzymic reactions also, the mechanism has been found to be almost similar. Thiamine cleaves the carbonyl - x bond, where x is generally —H or —COO^- group. A substrate-thiamine complex is later on cleaved at a-carbon atom to liberate the coenzyme.

Coenzyme A

This coenzyme is a complex thiol derivative and is usually written as CoA—SH but CoA is also in common usage. In contrast to coenzyme I and coenzyme II, it is not an oxidisingreducing coenzyme but it is acylating; it is an acyl transfer coenzyme and a cofactor for a wide variety of biologically acylations, e.g., the formation of acetoacetate from acetate in pigeon liver.

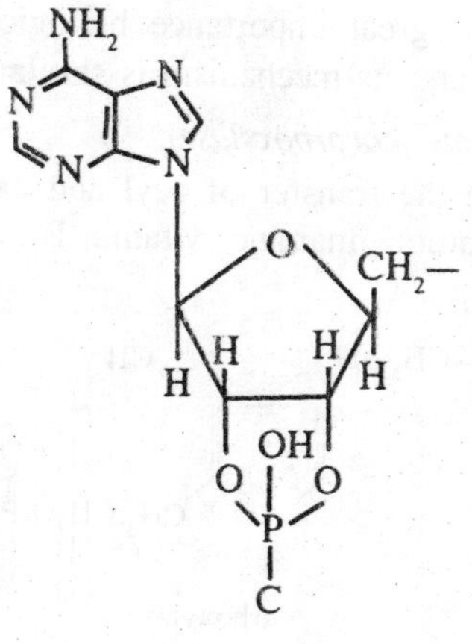

where **denotes the following group

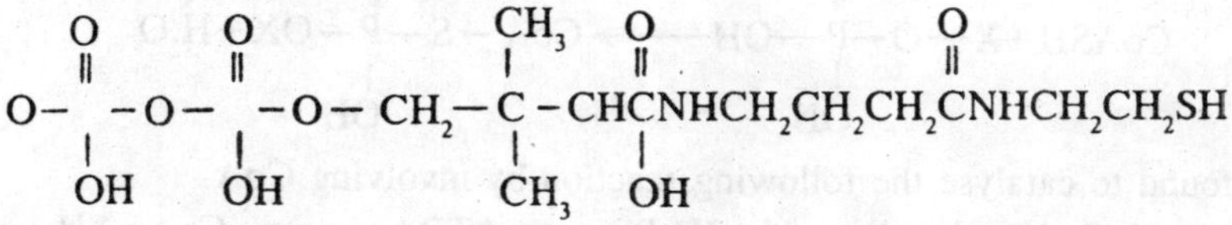

$$-O-\overset{O}{\underset{OH}{\overset{\|}{\underset{|}{P}}}}-O-\overset{O}{\underset{OH}{\overset{\|}{\underset{|}{P}}}}-O-CH_2-\overset{CH_3}{\underset{CH_3}{\overset{|}{\underset{|}{C}}}}-\underset{OH}{\underset{|}{CH}}\overset{O}{\overset{\|}{C}}NHCH_2CH_2\overset{O}{\overset{\|}{C}}NHCH_2CH_2SH$$

Role

Coenzyme A is required for the acetylation reactions. Coenzyme A accepts acetyl groups from one metabolite and donates them to another in the presence of specific enzymes.

The most important CoA compound is acetyl. CoA (activated acetate), which takes part in many acyl transferring reactions. In this, the acyl residue is bound to free —SH groups to form very reactive thioester.

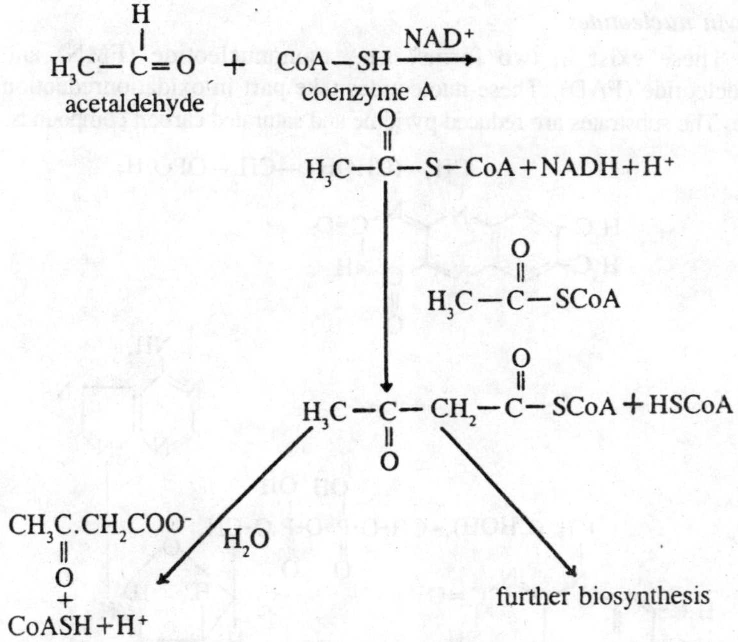

$$\underset{\text{acetaldehyde}}{H_3C-\overset{H}{\overset{|}{C}}=O} + \underset{\text{coenzyme A}}{CoA-SH} \xrightarrow{NAD^+} H_3C-\overset{O}{\overset{\|}{C}}-S-CoA+NADH+H^+$$

$$\downarrow \quad H_3C-\overset{O}{\overset{\|}{C}}-SCoA$$

$$H_3C-\underset{O}{\underset{\|}{C}}-CH_2-\overset{O}{\overset{\|}{C}}-SCoA+HSCoA$$

$$\xrightarrow{H_2O} CH_3\underset{O}{\underset{\|}{C}}.CH_2COO^- + CoASH+H^+$$

$$\longrightarrow \text{further biosynthesis}$$

In the formation of aceto acetyl CcA from acetate thioester there occurs nucleophilic displacement at the carbon and one more acetate group gets added at the place of coenzyme A.

Besides, acylation, coenzyme A can also undergo phosphorylation. In the reactions catalysed by phosphotransacetylase, thionase and phosphokinase, phosphorylation of CoA occurs. uokinases have been

$$CoASH + X{-}O{-}\overset{\overset{O}{\|}}{\underset{\underset{OH}{|}}{P}}{-}OH \longrightarrow CoA{-}S{-}\overset{\overset{O}{\|}}{\underset{\underset{OH}{|}}{P}}{-}OX + H_2O$$

found to catalyse the following reaction by involving CoA.

etyl CoA + adenylic acid + $2H_3PO_3$ $\rightarrow$ ATP + acetate + CoA—SH

Coenzyme A also takes part in the elimination reaction. For example in the formation of unsaturated fatty acids from saturated fatty acid coenzyme is involved. This needs another coenzyme FAD.

$$FAD + {-}C{-}C{-}\overset{\overset{O}{\|}}{C}{-}SCoA \rightarrow {-}\overset{|}{C}{=}\overset{|}{C}{-}\underset{\underset{SCoA}{|}}{\overset{\overset{O}{\|}}{C}} + FADH_2$$

Flavin nucleotides

These exist in two forms, i.e., mononucleotide (FMN) and dinucleotide (FAD). These nucleotides take part in oxidationreduction also. The substrates are reduced pyridine and saturated carbon compounds.

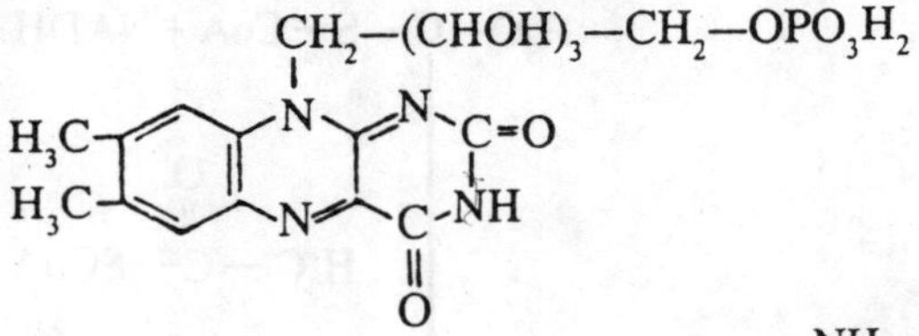

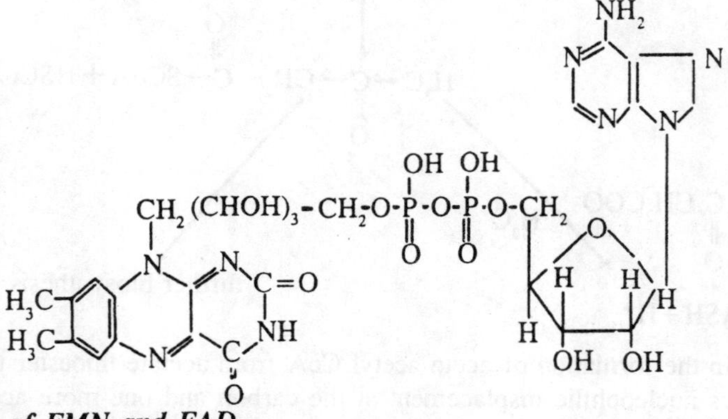

Role of FMN and FAD

Similar to NAD^+ and $NADP^+$, FMN and FAD are dehydrogenases. However, there is a distinction between NAD^+ and $NADP^+$ catalysis and FMN and FAD catalysis. NAD^+ (or $NADP^+$) carries hydrogen from substrate AH_2 to substrate B with no immediate need for oxygen; FMN and FAD, on the other hand, carry hydrogen from AH_2, to oxygen directly as shown schematically by means of curved arrows:

AH_2 NAD^+ BH_2

A NADH + H^+ B

AH_2 FMN H_2O_2

A $FMNH_2$ O_2

Curved arrows are used to represent the flow of hydrogen.

In FMN and FAD, the flavin is reduced. The reduction is accompanied by a loss of yellow colour. This colour is due to the system of conjugated double bonds seen in the oxidised form and lost upon reduction.

H_3C, H_3C, R, N, N, O, NH, O $+ 2H \rightleftharpoons$ H_3C, H_3C, R, N, H, N, O, NH, N, H, O

KINETIC PROPERTIES OF ENZYME SYSTEMS RELATED TO THEIR QUATERNARY STRUCTURE

Allosteric Enzymes

Although most enzyme activities have apparent temperature tptima when measured in vitro, the isothermal conditions characteristic of higher forms of life would preclude any major ontribution of temperature-dependent rate changes in the control of enzyme action. Slight changes in $[H^+]$ in the environnent of an enzyme, on the other hand, can have significant effect on activity. Often, the $[H^+]$-activity profile of an enzyme has the following shape:

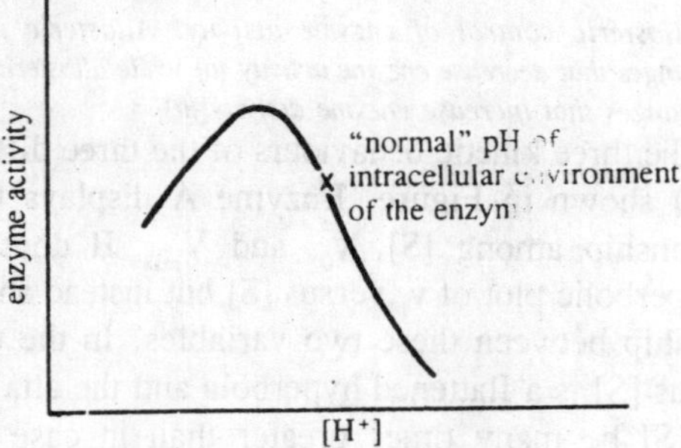

If the intracellular environment of the enzyme were on either the upslope or down-slope of the $[H^+]$-activity curve, it is reasonable that the activity could be sensitive to changes in $[H^+]$. Because they are macromolecular electrolytes, enzymes are also affected by ions such as K^+ and Mg^+. Such influences can contribute to cellular control of an enzyme's activity and, therefore, are relevant to metabolic regulation.

A more specific mode of regulatio of catalytic activity is achieved with enzymes that catalyse the first and the irreversible step in a multireaction sequence. These enzymes are oligomeric and respond extremely rapidly to either inhibitory or stimulatory modulators. Frequently, with such enzymes the substrate itself participates in the modulatory function. We shall analyse this type of regulation first.

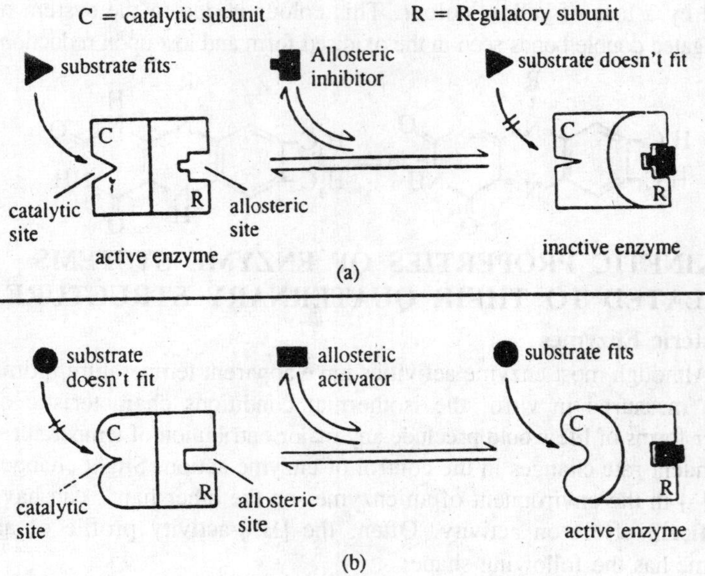

Figure 6.24 : Allosteric control of enzyme activity. Allosteric inl,ibitors induce conformational changes that decrease enzyme activity (a) while allosteric activators induce conformational changes that increase enzyme activity (b).

Consider the three kinetic behaviours of the three different enzymes (A, B, and C) shown in Figure. Enzyme A displays the Michaelis-Menten relationship among [S], v_0, and V_{max}. B does not give the rectangular hyperbolic plot of v_0 versus [S] but instead shows a sigmoid curve relationship between these two variables. In the case of C, the plot of v_0 versus [S] is a flattened hyperbola and the attainment of V_{max} requires that [S] be many times greater than in case A. It will be assumed that the value of V_{max} is the same for all three enzymes. Comparing enzymes A and B, it is seen that in the steep portion of the siginoid curve, the change in v_0 as a function of increasing [S] is many times greater than for the corresponding part of the hyperbolic curve. One conclusion that can be drawn is that when enzyme B begins to bind its substrate, combination with additional substrate is enhanced. The consequence of this acilitation of substrate binding is a marked acceleration in the v_0 of the enzyme reaction as evidenced by the sharp upturn in the sigmoid plot. This effect has been termed *positive cooperativity*. In the plot of v_0 versus [S] for C, it is apparent thar overall, the reaction rate with this enzyme is not as sensitive to small increases in [S] as in the case of A or B. V_{max} is approached much more slowly and at a much higher [S] than with A and B. These data are consistent with a *negative cooperativity* effect in which the binding of one substrate molecule makes the union and additional substrate more difficult.

As pointed out above, the double-reciprocal plots ($1/v_0$ versus l [S]) that are useful for analysing Michaelis-Menten kinetics are not applicable to enzyme systems showing either positive or negative cooperative behaviour. For example, the transposition of the curves to the reciprocal form would give the following plots:

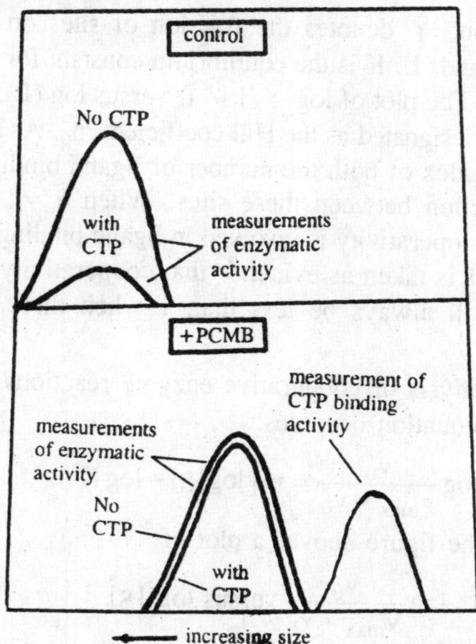

Figure 6.25: Experimental demonstration of the existence of separate catalytic and regulatory subunits in the allosteric enzyme l aspartate transcarbamoylase.

The most direct treatment of kinetic data such as those illustrated by curves (B) and (C) makes use of a logarithmic relationship, as shown below:

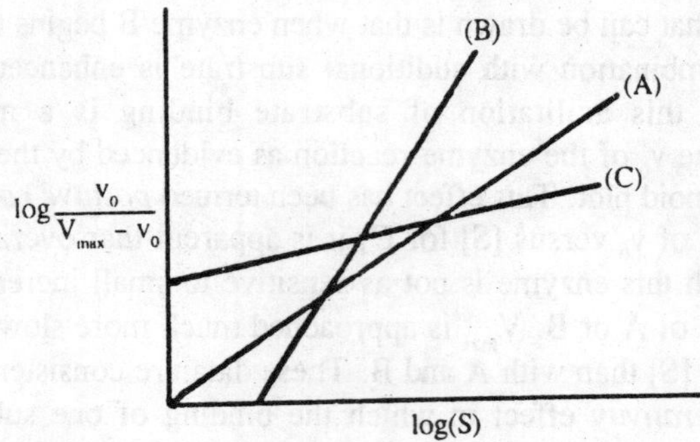

These plots are modeled after the ***hill plot***. Hill equation, which has been useful in describing the relationship between ligand binding by an oligomerie protein, can be formulated as:

$$\log \frac{\bar{Y}}{1-\bar{Y}} = n_H \log[L] - \log K$$

In this equation, $\bar{Y}$ denotes the fraction of sites on the protein occupied by the ligand, L. K is the equilibrium constant for the protein-ligand equilibrium. The plot of log $\bar{Y}/1 - \bar{Y}$ versus log (L) is a straight line whose slope is designated as the Hill coefficient, n_H. As an empirical constant, n_H is an index of both the number of ligand binding sites and the mode of intraction between these sites. When $n_H = 1$, it may be concluded that no cooperativity is involved in ligand binding. Any value of n_H greater than 1 is taken as evidence that cooperativity is positive; the value of n_H will always be less than 1 when there is negative cooperativity.

For kinetic analysis of cooperative enzyme reactions, the analog of the Hill ligand equation becomes:

$$\log \frac{v_0}{v_{max} - v_0} = n_H \log[S] - \log K$$

As shown in the figure above, a plot of

$$\log \frac{v_0}{V_{max} - v_0} \text{ versus } \log [S]$$

will give a straight line whose slope is the value of n_H. In applying the Hill equation to enzyme kinetic data, two assumptions are made; V_{max} is attained when all of the substrate bind1ng sites are occupied and the rate-limiting step in the entire reaction sequence is the binding of S by the enzyme.

The kinetic data presented in the figure illustrate how the activities of allosteric enzymes can be modulated or regulated solely by their substrates and that the degree of modulation is determined by the number of the multiple-substrate binding sites occupied. Allosteric enzymes can also be modulated by molecules other then their substrates and such regulation may be expressed either as stimulation or as an inhibition of activity.

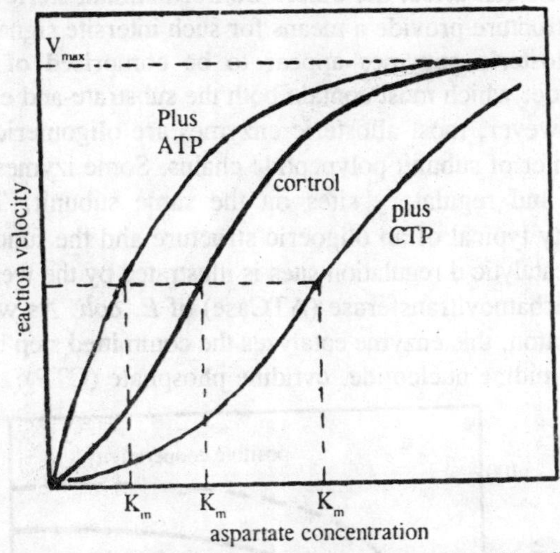

Figure 6.26: Allosteric regulation of aspartate trascarbamoylase. The all steric inhibitor CTP (increasess the K_m of the reaction while the allosteric activator decreases the K_m.

Their ability to respond to modulator molecules other than their substrates is the central reason for calling these enzymes *allosteric*. The term "*allosteric*" means "another place or structure" and aptly describes these regulatory enzymes which have been shown to have modulator-binding sites, in addition to the catalytic substrate-binding sites. Like the substrate-binding site(s), the allosteric site(s) can exhibit a high degree of specificity in binding a modulator molecule.

Modulators be positive (stimulatory) or negative (inhibitory) and the response of the affected enzymes will be reflected in the plots of v_0 versus [S] as shown in the figure. Of two types of allosteric enzymes described here, (a) is one in which S binding is enhanced by positive cooperativity, and (b) is an enzyme in which negative cooperativity hinders S binding. In contrast to enzyme reactions showing Michaelis-Menten kinetic relationship between v_0 and [S], the K_M values for allosteric enzymes must be labeled as "apparent K_M" values and are equated to the [S] values giving half-maximal velocity. The positive and negative modulator effects on the apparent K_M and V_{max} values of the two enzymes described in the figure, are summerised by the following generalisations.

With either negative or positive modulator of enzyme activity, the catalytic and regulator sites must have some facility for intercommunication, either direct or indirect, by which changes occurring in one binding site can affect the other. Conformational steric changes in the protein structure provide a means for such intersite signal transmission. Some allosteric enzymes appear to be comprised of only a single polypeptide, which must contain both the substrate-and effector-binding sites. However, most allosteric enzymes are oligomeric and intain an even num er of subunit polypeptide chains. Some izymes have both the catalytic and regulatory sites on the same subunit. The molecular complexity typical of an oligoeric structure and the functional relation between catalytic d regulation sites is illustrated by the well characterised partate carbamoyltransferase (ATCase) of *E. coli*. As will be tailed in a later section, this enzyme catalyzes the committed step in the synthesis of the primidine nucleotide, cytidine phosphate (CTP):

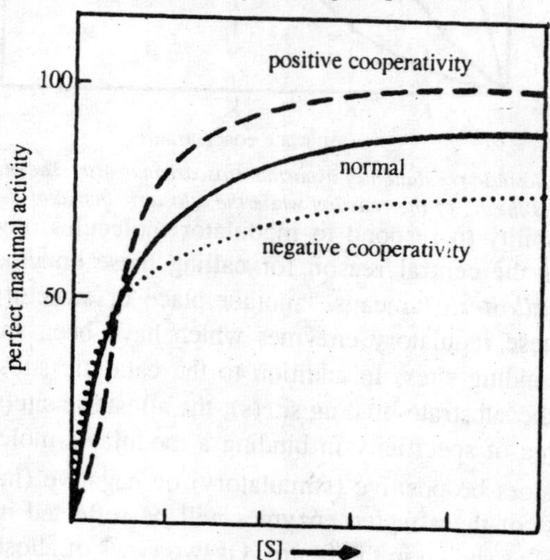

Figure 6 27: Effects of negative and positive cooperativity on enzyme kinetics.

$$\text{L-Asparate} + \text{carbamoyl phosphate} \xrightarrow[\text{seven step}]{\text{ATCase}} \text{carbamoyl}$$

asparate →→→→→→→→ CTP →→→ nucleic acids

The allosteric character of ATCase is illustrated by the kinetic data. As with any enzymatic reaction involving two substrates, the initial rates shown here be measured under conditions in which the concentration of only one of the, substrates is made the variable. In such analyses, the ATCase is saturated with carbatnoyl phosphate and v_o thus becomes dependent on the aspartate concentration.

As ATCase is an enzyme that exhibits a positive cooperativity in the binding of substrate. This enzyme also presents a model system for the demonstration of both positive and negative modulation. In the presence of ATP, a positive modulator of ATCase, the apparent Km is decreased and the initial rates are higher than in the absence of ATP. The converse is found when CTP is added to the system. Acting as atypical negative modulator, the CTP can bind to the enzymes and causes a decrease in activity and an increase in the apparent Km.

The knetic behaviour of ATCase has been found to be sensitive to the quaternary structure of the enzyme. The *E. coli* enzym : for example, can be dissociated into two identical catalytic subunits (each containing three polypeptide chains) and three identical regulatory subunits (each containing two peptide chains). Each catalytic subunit contains three substrate binding sites, persumably one on each peptide chain.

Correspondingly, each regulatory subunit has two binding sites, apparently one per peptide chain. Evidence for interaction between these 12 binding sites, can only be obtained with the intact ATCase. Thus, the catalytic subunits alone exhibit more of a Michaelis-Menten kinetic type of behaviour than positive cooperativity in subtrate binding. Similarly, activity of the catalytic subunits is not affected by the addition of CTP. Modulatory control by CTP is only observed when the catalytic and regulatory subunits are physically associated as permitted by the oligomeric form of the enzyme.

All models proposed for relating kinetic behaviour and macromolecular structure of oligomeric allosteric enzymes assume that both the catalytic and regulatory subunits can have two conformational states in equilibrium mixture: one, which is inactive, with a low affinity for S, and the other active, with a high affinity for S.

The scheme given in figure illustrates the interconversion of these forms in positive cooperative binding of S and allosteric regulation by positive and negative modulators. As shown in figure when substrate combines with the catalytic site, the enzyme is "locked" into the active form, causing a shift in equlilbrium of the two states as well as

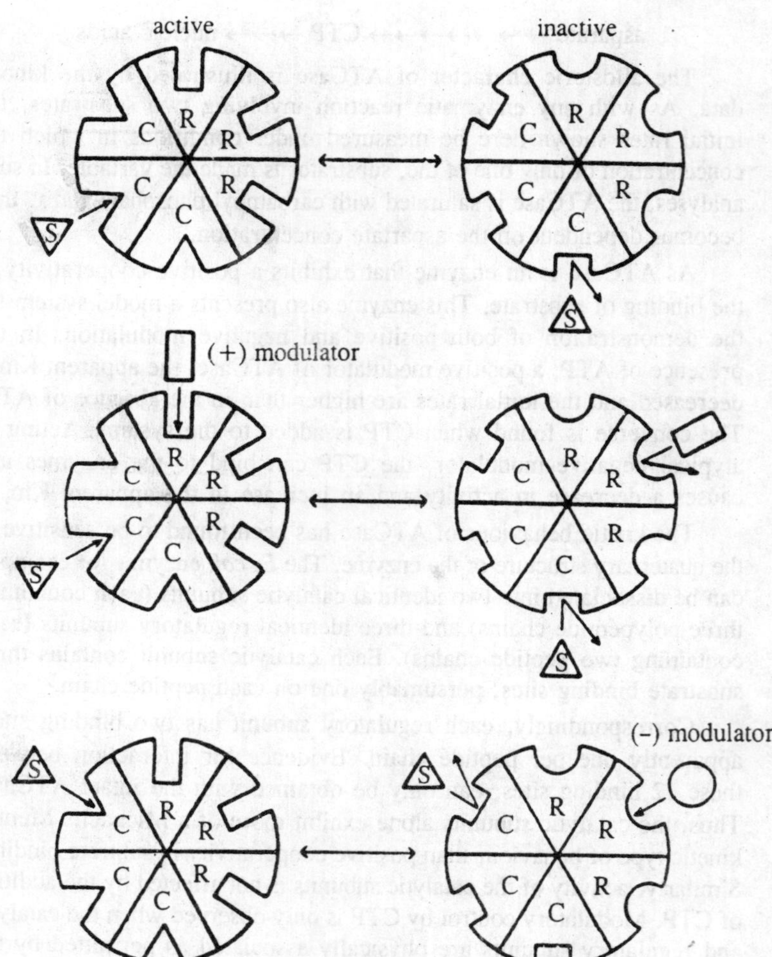

Figure 6.28: Conformational states of allosteric enzymes.

facilitating union of S with the remaining site on active forms. This model also describes the role of the regulatory subunits in the modulation of activity of the oligomeric enzyme. The binding site on the regulatory subunit of the activity form of the enzyme has the proper configuration for binding a positive modulator. The direct consequence of this binding is to faci-litate locking of the enzyme in the active form. In contrast to the accelerating effect of a positive modulator, binding of an inhibitory modulator by the regulatory subunits is favoured by the enzyme's inactive form'and, therefore, will stabilize this conformation. The initial rate of the enzyme reaction is inhibited accordingly.

The contribution of enzymes to metabolic regulation will be treated in a broader context is a later, scction. In the present discussion devoted primarily to the molecular basis of such regulatory mechanisms, the principal focus' has been on the peculiar changes in conformation of some enzymes and how such changes have the potential for fine tuning of metabolic controls. The inextricable relationship between structure and function that is so beautifully illustrated by the allosteric enzymes can also be expressed in other ways. Two additional examples of nature's versatility in enzymatic regulation are (1) the enzymes whose activity depends on covalent modifica-tion and (2) the isozymes.

Enzymes that can be modified covalently

Covalent modification may take several forms. In the case of the less active form of phosphorylase (phosphorylase *b*), the enzymes responsible for catalyzing the conversion of glycogen to glucose-l-phosphate, serine residues are phosphorylated enzymatically to convert the enzyme b to its active oilgomer, enzyme *a*: Another example of covalent modification is the formation of hydrolytic enzymes from inactive precursors.

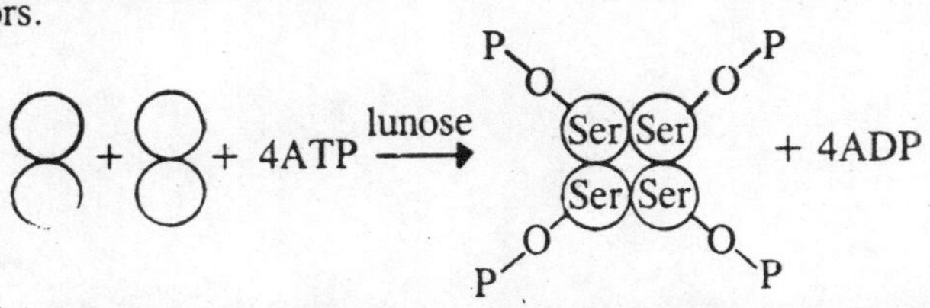

P–O, O–P

+ + 4ATP $\xrightarrow{\text{lunose}}$ Ser Ser Ser Ser + 4ADP

P–O, O–P

These enzymes are essential for digestion of proteins in the gastroint-estinal tract and are synthesized and secreted in catalytically inactive forms designated as zymogens. The hydrolytic rupture of specific peptide bonds in the zymogens renders them enzymatically active.

$$\text{zymogen} \xrightarrow[\text{enzymatic}]{\text{HOH}} \text{active enzyme} + \text{smaller peptide(s)}$$

The details of this activation process and its physiologic implications for digestion are treated later.

Isozymes

First recognised by the powerful analytical tool of electrophoresis, the isozymes provide another example of low biologic processes can endowed with sensititive controls because of the functional variations which become possible when polypeptide chains or subunits of an oligomeric enzyme are harmoniously associated, Lactate dehydrogenase, the terminal catalyst in glycolytic cenvension of glucose to lactate, is such an enzyme.

Characterisation of the lactate debydrogenase of rat tissues has demonstrated that the enzyme can occur in as many as five isozymic forms. Each of these enzymes contains four polypeptide chains. These polypeptide subunits have been shown to be of only two kinds and are commonly labelled as the H (for heart) and the M (for muscle) chains. The single chains are inactive. The five tetrameric isozymes of lactate debydrogenase are the five different combinations of these subunits

HHHH HHHM HHMM HMMM MMMM

The *Km* and *Vmax*' values and the degree of product inhibition of the nemologous tetramers differ significantly. The advantages of these differences in knetic properties of the lactate debydrogenase isozymes are described in a later section which treats regulation of metabolic activity in the heart and skeletal muscle.

7

Biomanufacturing of Protein

The proteins are highly organised and complex structures. The proteins have significant role in the structural and functional organisation of the cell. The structural proteins constitute various cellular components and some extracellular parts such as cuticle and fibres, etc. The functional (e.g., enzymatic and hormonal) proteins control almost all metabolic, *biosynthetic*, *bioenergetic*, *growth regulating*, *sensory* and *reproductive* activities of the cell.

All the proteins which are required by the cell for its different purposes are synthesized by the cell itself intracellularly. Chemically, proteins are polymers of amino acids. The amino acids molecules are linked together in linear fashion forming long polypeptide chain. The adjacent amino acid are joined together by covalent peptide bond between the *carboxy group* of one amino acid and the amino group of the other.

DNA initiates the mechanism since it carries the genetic message for protein synthesis. This genetic message from one DNA strand is transferred to a complementary strand of messenger RNA ; mRNA then passes out of the nucleus and carry the message to the ribosomes where it is used to synthesize the particular type of protein molecule.

The first step is called *transcription* because it involves copying of the genetic information from DNA into mRNA which transmits the coded information from DNA to ribosomes, the site of protein synthesis. The second step is termed as *translation* as it involves the translation of the information encoded on mRNA with newly synthesized protein *molecule*.

CENTRAL DOGMA AND REVERSE CENTRAL DOGMA

It is a sort of relationship between DNA, RNA and protein. It is details with the flow of genetic information and can be studied under the following heads:

One way flow of informations

Crick suggested that DNA transmits the genetic informations for protein synthesis by transcribing mRNA (*transcription*) and other types of RNAs. These RNAs then translate the nucleotide sequence into the sequence of amino acids (*translation*). This is known as one way flow of information or central dogma.

Circular flow of information

Barry Commoner suggested circular flow of information. The DNA transcribes RNA (mRNA, tRNA etc.), RNA translates the information into protein, protein then directs the svnthesis of RNA and RNA dictates the synthesis of DNA.

Inverse flow of information

Temin has described the presence of an enzyme-RNA, *dependent DNA polymerase,* which controls the synthesis of DNA from a single strand of RNA. *D. Baltimore* has described the presence of this enzyme in certain RNA tumour viruses. In tumour viruses the RNA is genetic RNA but in certain cases the nongenetic RNA synthesized under the control of DNA may also control the synthesis of DNA by inverse flow of information. This is known as central dogma reverse.

Role of Amino Acids

Amino acids are the units from which the protein molecules :are made. In the cytoplasm of every cell a number of amino acids are present. Out of these 20 amino acids take the important part in protein synthesis. The amino acids are distributed through out the *cytoplasm*, which are collected by specific tRNAs. tRNAs bring the amino acids to the actual site of protein synthesis i.e., to ribosomes where the synthesis of protein takes place.

ROLE CF mRNA DURING SYNTHESIS OF PROTEIN

According to *Jacob* and *Monod* mRNA conveys necessary message of genetic code from the DNA of the nucleus to the cytoplasm. It acts as template for the translation of the DNA code into specific protein. It is formed from DNA during *replication* with similar base composition.

The evidence that DNA is the main source from which RNA is synthesized can be received by using *deoxyribonuclease*, which ceases protein synthesis.

The inhibition of protein synthesis is due to a fact, that mRNA is not formed in the absence of DNA as it is digested by the enzyme. *Hall* further gave similar evidence regarding the formation of mRNA from DNA. Messenger RNA molecules necessarily exist in a large variety of length, depending upon the length of the polypeptide chain for which they code. There are very few polypeptide chains which contain less than 100 amino acids, and so very few molecules which contain less than 300 nucleotide units, the length of mRNA which carries information necessary to determine the complete polypeptide chain of a protein molecule is called a *cistrof*, as postulated by *Benzer*.

Some mRNA molecules are able to carry information for the synthesis of more than one protein molecule. Such types are known as polycistronic. The average life time of mRNA molecules is variable. In bacterial cell mRNA molecules exist for few minutes at 37°C, after which they are broken down into its components by enzymes. In the case of *mammalian reticulocytes* the mRNA are long lived.

There is then considerable variation in average lifetimes of mRNA molecules. While studying the impact of radioactive substances on the incubated cells, it has been observed that the labelled material is first incorporated into the nuclear RNA and then appears in the cytoplasmic RNA. This indicates that when the code has been transcribed from DNA into mRNA, the latter leaves the nucleus. In the cytoplasm it reaches to the site of protein synthesis, the ribosome.

The mRNA molecule attaches reversibly to the surface of the ribosome. The attachment is ***manifested*** by Mg^{++} (magnesium cation) and finally a complex is formed, between mRNA and the ribosome.

Role of tRNA

It is also called as soluble RNA. This particular nucleic acid is present in the cytoplasm in sufficient amount and the main pole of tRNA is to pick up the amino acid from cytoplasm and bring to the site of protein synthesis on the ribosome. As already stated that a number of amino acids (20) are present in the cytoplasm and for every specific amino acid there is one specific t-RNA, so there are different specific tRNA molecules.

A specific adopter enzymes (an-amino-acyl tRNA synthetase) is required to couple each amino acid to its specific tRNA. The part of tRNA which becomes attached to its particular amino acid appears to be same for all sorts of tRNAs. This limb is the 3' end of the molecule, terminating in the sequence of three base-*Cytosine-Cytosine-Adenine*. Another 5' end of the molecule terminates in the base *guanine*. The sequence only enables amino acid to be attached at particular site.

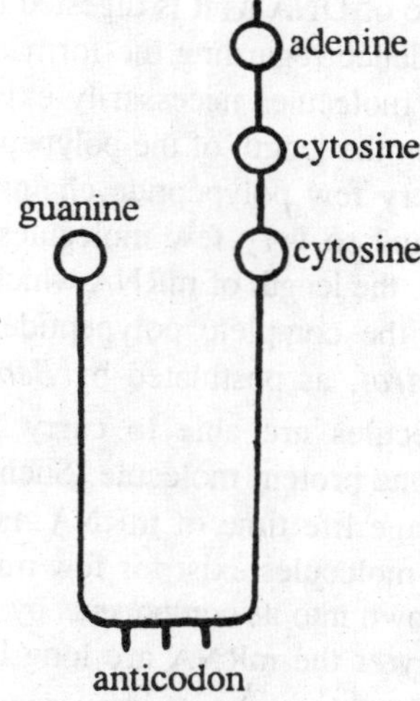

Figure 7.1 : Sinxture of tRNA molecule.

The process by which the amino acids become attached to the tRNA is quite complicated. ATP provides energy which is required during the process. The link formed between the amino acid and transfer RNA is a covalent bond between the carboxyl group of the amino acid, and the ribosome of the terminal adenine nucleotide of the tRNA. Thus there are three components necessary for the reaction

(*i*) a specific amino acid

(*ii*) a specific enzyme, and

(*iii*) a specific tRNA.

The reaction can be shown by the following equations

(*i*) Amino acid + ATP + Amino acyl adenylate → Enzyme bond Amino acid adenylate + 2P

(*ii*) Enzyme bound Amino acid adenylate + uncharged tRNA → charged -RNA + Enzyme + amino acid complex.

POLYRIBOSOMES AND THEIR ROLE IN PROTEIN SYNTHESIS

The 70S free ribosomes are the active units in protein synthesis in a normal cell. The most active units are aggregates containing a mRNA molecule to which a number of 70S ribosomes are attached. The name polyribosome is designated to this assembly line of ribosomes. Before the study of structure of *polyribosomes* let us know the history and the conception by which the idea of polyribosomes appeared. *Alexander Rich* studied and expressed in his words "consider for a moment the problem of synthesizing one of the polypeptide chain of, *haemoglobin*, which contains about 150 amino acid sub-units. If each such-unit is specified by a triplet code, the m-RNA must contain 450 bases merely to specify the sequence of sub-units in one chain. In most RNA molecules the bases are *stacked* on top of one another more or less like a pilo of pennie. Since the bases have a thickness of 3.4 Å units, the mRNA strand for the *haemoglobin* polypeptide chains should have a molecular length of at least 1,500Å. In other possible arrangements of the RNA molecule the length might be almost twice By comparison, the individual ribosome has a diameter of only about 230Å, and so we wondered how the long messenger molecule interacted with such a small particle to manufacture a *polypeptide chain*".

Some investigators thought that the RNA chain might be wrapped around the out side of the ribosome, but it was bard to visualise bow intimate contract between, the two should be maintained. The wrapping problem would be still more difficult for RNA chain 20,000Å long, which are found in many viruses. Alternative suggestions that the mRNA might come how be coiled inside the ribosomes seemed to present even more formidable *tonographical* problems.

If occurred to us that proteins might actually be made on groups of ribosomes lined together some how by mRNA. There was already a little evidence pointing in this direction. *Walter Gilbert* of Havard University, as well as other investigators, had found that when a synthetic RNA was added to a cell free system of *bacterial ribosomes* the ribosomes would tend to clump together. Probably the first polyribosomes were studied in *mammalian* (rabbit) *reticulocytes* that *predominantly* synthesise the blood protein haemoglobin contains ribosomes.

The negative staining procedure of *Slayter* shows that ribosomes in a polyribosome are separated by gaps of 50 to 150Å while positive staining with uranyl acetate reveals that the ribosomes are connected by a thin thread 10 to 151 in diameter which is about the thickness of a single strand of RNA.

From these studies, it is concluded that the over all length covered by a five unit polyribosomes is about 1,500Å. The space covered by the polyribosome is nearly the same which is needed for mRNA strand in the synthesis of haemoglobin in reticulocytes.

POLYRIBOSOME CYCLE AND ASSOCIATION AND DISSOCIATION OF SUBUNITS

Few years ago it was thought that the various molecules involved in the protein synthesis were bound on the complete, ribosome. However, *Mangiarotti* and *Sclessinger* have shown that the two sub-units of the ribosome may exist freely and that they associate during the protein synthesis and may dissociate again at the end of the process.

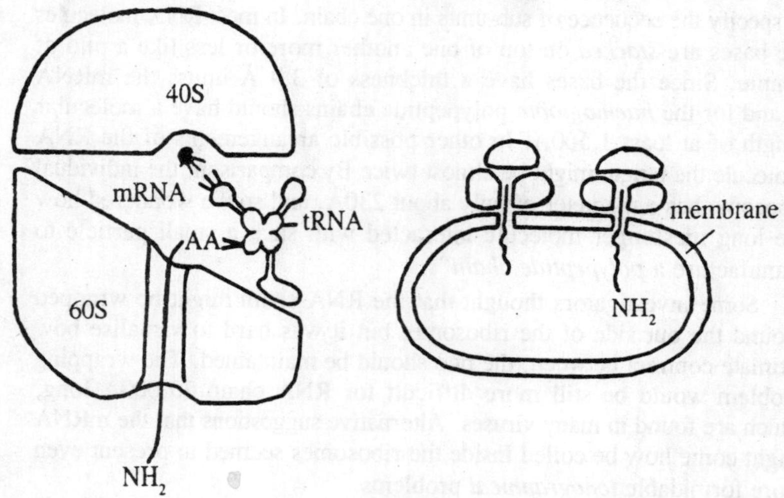

Figure 7.2: Diagram of ribosome showing the two sub-units and probable position of mRNA and tRNA.

Kalmpfer and his associates have shown the essential features of ribosomal cycle as indicated in the figure. In this figure, the two different subunits (30S and 50S in *E. coli* or 40S and 60S in reticulocytes) enter in the process at the initiator end and *separate* at the *terminator* end after the completion of the cycle.

Thus, according to this concept the subunits of different ribosomes- could be exchanged during the *ribosomal cycle*. The factors which are responsible for the initiation of the 30S particles in the ribosomal cycle, are identified as F_1, F_2, F_3 + ATP. The separation factor is, however, tentatively identified as F_2.

POLE OF SUBUNITS IN PROTEIN SYNTHESIS

These subunits (30S and 50S or 40S and 60S) perform. different types of functions in the process of protein synthesis. In *E. coli*, mRNA binds to the KS subunit for the reading of genetic message. In first step small 30S sub-unit binds with the first codon of the mRNA to form the so-called *initiator* complex. The successive steps would be the binding of the first amino acyl-tRNA and then the coupling of the other subunit (50S) as shown in the figure elsewhere in this chapter.

In *E. coil* it has been demonstrated that the first aminoacyl tRNA which initiates the synthesis of protein is quite specific because it contains a formyl (-CHO) derivative of *methionine* which is coded by AUG codon. This codon is always present at the beginning of all the mRNA molecules. In the protein synthesis the peptide chain always grows from the free terminal end of tNH, group towards the -COON end.

Therefore, the function of *formyl methionine* tRNA (F-met -tRNA of the figure) is to ensure that proteins are synthesized in that direction. In F-met-tRNA, the NH, group is blocked by the *formyl group* leaving only the -COOH available to react with the -NH, of the second amino acid. Thus, in this way the synthesis follows in only correct direction and sequence.

The 50S subunits contains the enzyme peptidyl transferase or peptide synthetase, which is essential in the formation of peptide bond. The other function of the 50S subunit is to provide two binding sites for the two tRNA molecules. These two sites should be next to each other in order to permit the formation of the peptide bond. However, it is possible to envision process of protein synthesis as a continual entrance of *aminoacyl* tRNA on 30S unit of the ribosome and exist of the tRNA molecules from the 50S sub unit after giving their aminoacids to the developing *polypeptide chain*.

MECHANISM OF PROTEIN SYNTHESIS

The mechanism of protein synthesis has been investigated most thoroughly in bacteria (usually *E. coli)*, therefore, the process of protein synthesis of *E. coil* which contains 70S ribosomes will be described in more detail and then this will be discussed in relation to our present knowledge of protein. synthesis on 80S ribosomes of plants and animals.

Protein synthesis consists of two main events:

(*i*) Transcription

(*ii*) Translation

Transcription

The formation of m-RNA is called transcription. Daring, the process an enzyme-RNA polymerase is needed. The enzyme is then attached to the DNA at the initiation site, where a short segment of the DNA molecule is opened up. The complementary nucleotides developed from one of the DNA strands, form mRNA. The RNA polymerase moves along the DNA strands, and the growing m-RNA strands is *peeled* off till a *termination* site is reached.

The enzyme RNA polymerase consists of five different polypeptide chains, of which four collectively form the core enzyme while the fifth is called the sigma (*a*) chain. They are given below.

Core enzyme

(*a*) β having mol. weight 1,60,000

(*b*) β having mol. wieght 1,50,000

(*c*) α having mol. weight 40,000

(*d*) α having mol. weight 10,000

Sigma factor

(*a*) having mol. wt. 90,000.

The sigma chain recognises start signals along the DNA. molecules. A specific protein called the rho (ρ) factor (mot. wt. 60,000) stops the formation of the RNA chain. OnceRNA synthesis is initiated, the sigma factors dissociate and core enzymes bring about elongation of mRNA.

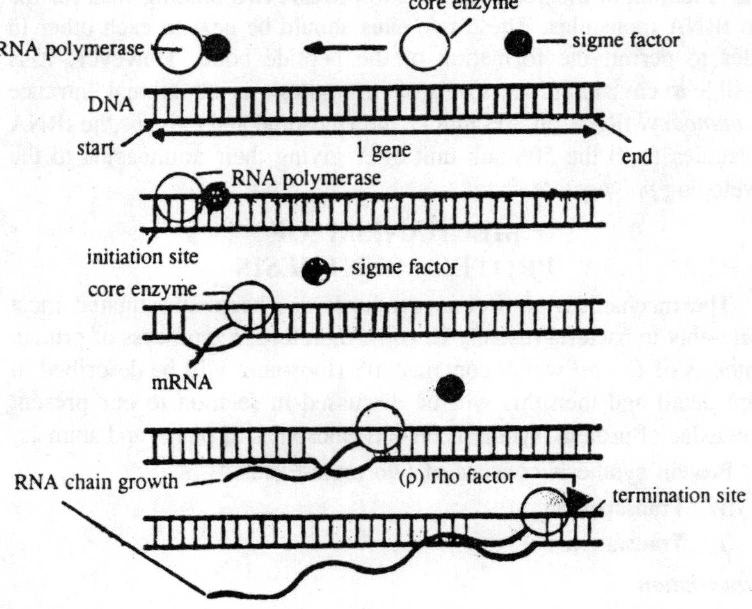

Figure 7.3 : Showing transcription.

The mRNA makes its way through the pores in the nuclear membrane to the cytoplasm to start the next stage which is translation.

Transcription in Prokaryotes

The important features of transcription in both prokaryotes and eukaryotes are

1. The product of transcription is single-stranded RNA.
2. For transcription to occur, the DNA must unwind locally. For each gene, only one of the two strands serves as the template strand for RNA synthesis : this strand is called the sense strand.
3. The RNA precursors (the molecules polymerised into the RNA chain) are the ribonucleoside triphosphates ATP, GTP, CTP, and UTP, collectively called NTPs. The polymerisation reaction is catalysed by the enzyme RNA *polymerase*.

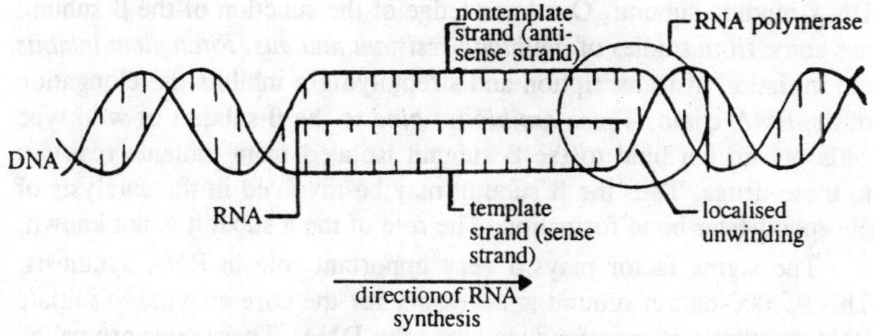

Figure 7.4: Schematic of the transcription of RNA from a DNA double helix.

4. The RNA polymerisation reaction is similar to the DNA polymerisation reaction: the next ribonucleotide to be added to the chain is selected by the RNA polymerase for its ability to form a complementary base pair with the exposed DNA base on the template strand.
5. RNA is synthesized in the 5' -to 3' direction. For example, if the template strand is 3'-TCGGAT-5', the RNA chain will be 5'-AGCCUA-3'.

Prokaryotic RNA Polymerases

Each prokaryotic organism has its own specific RNA poly merase. *Bacteriophages* either use the host *bacterium's* RNA polymerase or code for their own, depending upon the bacterio phage.

RNA polymerases require a double-stranded DNA template, the four *ribonucleoside triphosphates*, and *magnesium* ions to catalyse RNA synthesis. Unlike prokaryotic DNA polymerases, RNA polymerases are able to initiate new RNA chains, and they lack *proofreading* functions.

Among prokaryotic RNA polymerases, the enzyme from *E. coli* has been studied most extensively. This enzyme has a sedimentation coefficient of 11—13 S and a molecular weight of approximately 500,000 daltons. Relatively gentle treatment of the soluble enzyme dissociates a 95,000-dalton polypeptide called the sigma factor (σ), leaving the *core polymerase*. The latter can be further dissociated into four polypeptide chains. These are the *beta-prime* (β') subunit (165,000 daltons), *beta* (β) subunit (155,000 daltons), and two copies of the *alpha* (α) subunit (41,000 each).

The *core enzyme* alone can make an RNA copy of DNA, and the roles of the subunits in this process have been studied extensively. Isolated β' subunits, for example, have been shown to bind to DNA in vitro, whereas the a and β subunits do not. Thus apparently β' is the DNA-binding subunit. Our knowledge of the function of the β subunit has come from studies of *antibiotic-resistant mutants*. *Rifampicin inhibits* the initiation of transcription and streptolydigin inhibits the elongation of the RNA chain. These *antibiotics bind* to the β subunit of wild type cells but do not bind to the β subunit isolated from mutants resistant to these drugs. Thus the β subunit may be involved in the catalysis of phosphodiester bond formation. The role of the a subunit is not known.

The sigma factor plays a very important role in RNA synthesis. This 95,000-dalton subunit is necessary for the core enzyme to initiate RNA synthesis at specific sites along the DNA. These sites are called promoter sequences, or more simply, *promoters*. If a is not bound to the core enzyme. The latter initiates RNA synthesis at random sites along the DNA and, moreover, both strands are copied instead of one. The promoter recognition and transcription events catalysed by RNA polymerase occur very rapidly, often within 0.2 second.

The key process is the recognition of the promoter sequence in the DNA by the enzyme. Prokaryotic promoters that are recognised by the *E. coli* RNA polymerase have in common (with minor variations) a seven-nucleotide sequence called a *Pribnow box* found about five or six bases upstream of the actual start of transcription. The Pribnow box sequence on the anti-template strand has the general formula 5'-TAT purine AT purine-3'.

A second sequence called a *recognition sequence* is in a region about 35 base pairs upstream from the start of the RNA transcript. The RNA polymerase probably first binds there and then slides to the

Pribnow box to initiate transcription. After RNA synthesis has begun, the a factor dissociates from the core enzyme, which then continues the transcription process. The a factor may then associate with another core enzyme to initiate another RNA transcription at the same or a different promoter. Thus the a factor acts catalytically in terms of the initiation of transcription at specific sites.

There is evidence that the core RNA polymerase undergoes conformational changes. Specifically, its sensitivity to salt, proteases (*protein degrading enzymes*), and inhibitors changes when it becomes bound to DNA and when RNA synthesis is being catalysed. And, as was noted before, the enzyme is no longer sensitive to the RNA synthesis *inhibitor*, *rifampicin*, once chain initiation has begun.

When a gene has been transcribed, the RNA polymerase is released, and the newly synthesized RNA is displaced by the reformation of double-helical DNA. Transcription stops when a *transcription termination sequence* after a gene is recognised by a complex of the RNA polymerase core enzyme and a protein factor coded for by the *nusA* gene. When transcription is correctly. terminated, the *nusA* factor dissociates from the core enzyme and can be used for other termination reaction.

Initiation, Elongation, and Termination Steps of Transcription

Three types of base-pair sequence are needed for a gene to be transcribed : the DNA sequences that code for the base sequence of the RNA transcript (the gene), a sequence that signals the start point for transcription (the *promoter sequence*), and a sequence that signals the stopping point for transcription (the *terminator sequence*), 5'-promoter-gene-terminator-3'. The promoter is considered to be "*upstream*" from the gene and the terminator is considered to be "*downstream*".

Figure shows the general processes of initiation, elongation, -and termination of transcription. The important features are

1. RNA polymerase core enzyme plus a factor recognises and binds to the promoter sequence.
2. RNA polymerase slides toward the gene and catalyses the localised *denaturation* of the two DNA strands, exposing the template strand for the initiation of RNA synthesis.
3. RNA synthesis commences and the RNA chain is made in the 5'-to-3' direction. Once a few *polymerisation* reactions have been completed, the a factor is released.
4. The transcription termination sequence signals the RNA-

polymerase to cease transcribing the DNA. The terminator sequence consists of two parts: (a) a region with a high proportion of GC base parts (a GC-rich region) and (b) a series of T bases on the *non-template* strand in an AT-rich region. Both= of these regions are transcribed by RNA polymerase, and base pairing between two parts of the transcript in this region causes an altered relationship between the RNA polymerase and the DNA. As a result, the *RNA polymerase* cannot elongate the DNA chain any further, and it and the RNA chain both dissociate from the DNA. The termination reaction involves a complex between the RNA-polymerase and the *nus A*. A protein factor, and the release of the RNA transcript from, the DNA requires the activity of a ρ (*rho*) protein factor.

Transcription in Eukaryotes

Transcription in eukaryotes occurs in a very similar way to transcription in *prokaryotes*. The differences are that the transcription controlling sequences are different, and more than one RNA *polymerase* enzyme is used in eukaryotic organisms.

Eukaryotic RNA Polymerases

Three types of nuclear RNA polymerases (I, II, and III) have been found in a number of eukaryotic cells. These have been shown to have distinct specialised roles in RNA *transcription*, and they vary in their sensitivity to the RNA synthesis. inhibitor α-amanitin. None is sensitive to the prokaryotic RNA synthesis *inhibitor rifampicin*.

RNA Polymerase I

This enzyme is not inhibited at all by α-amanitin. It is found within the nucleolus and is responsible for the transcription of the genes for the 2' S, 5.8S, and 18S rRNA molecules.

RNA Polymerase II

This polymerase is inhibited at low concentrations of α-amanitin and is required for the synthesis of most other RNA species, notably mRNA.

RNA Polymerase III

This enzyme is inhibited at high concentrations of α-amanitin and functions in the transcription of tRNA (4S RNA) and 5S ribosomal RNA.

All of these RNA polymerases are complex, each consisting of a

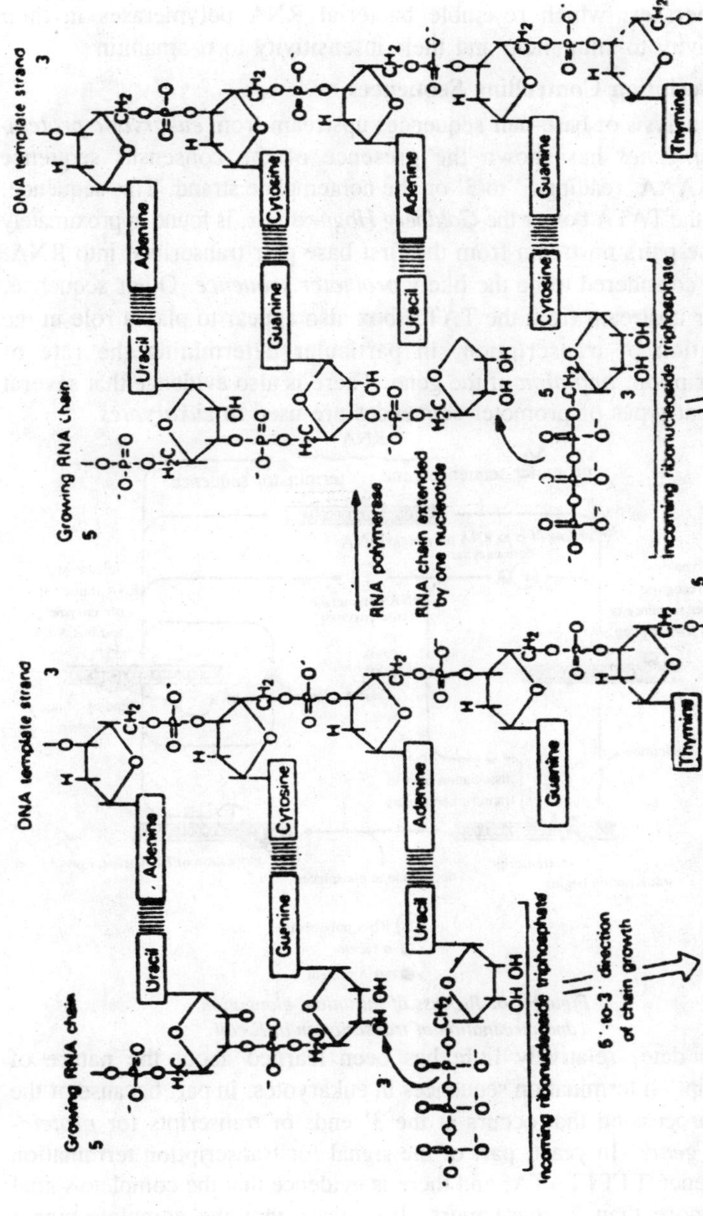

Figure 7.5: Chemical reaction involved in the RNA polymerase-catalysed synthesis of RNA on a DNA template.

number of polypeptide subunits.

The mitochondria and chloroplasts contain their own RNA polymerases, which resemble bacterial RNA polymerases in their sensitivity to rifampicin and their insensitivity to α-amanitin.

Transcription Controlling Sequences

Analysis of base-pair sequences upstream from *eukaryotic protein-coding genes* has shown the presence of the consensus sequence TATAAAA, reading 5'-to-3' on the nontemplate strand. This sequence, called the TATA box or the *Goldberg Hogness box*, is found approximately 30 base pairs upstream from the first base pair transcribed into RNA, and is considered to be the likely *promoter sequence*. Other sequence, further upstream from the TATA box also appear to play a role in the regulation of transcription, in particular determining the rate of transcription, *initiation* of the gene. There is also evidence that several different types of promoter sequences are used in *eukaryotes*.

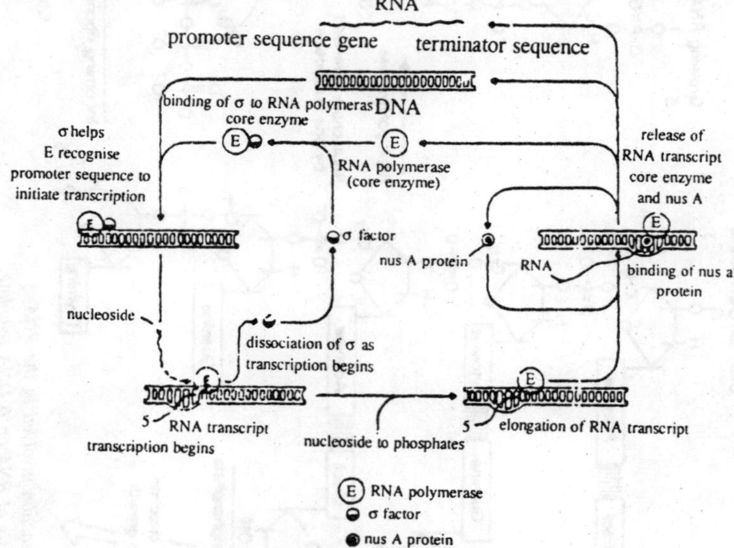

Figure 7.6: Process of initiation, elongation, and termination of transcription in E.coli.

To date, relatively little has been learned about the nature of transcription termination sequences in eukaryotes: in part because of the rapid processing that occurs at the 3' ends of transcripts for *protein-coding genes*. In yeast, part of the signal for transcription termination is sequence TTTTTATA, and there is evidence that the complete signal is no more than 21 base pairs. It appears that the complete signal requires the presence of other sequences *downstream* for full effectiveness.

Messenger RNA

General properties

Genes coding for proteins are transcribed into mRNA *molecules*, which become associated with ribosomes where protein synthesis occurs. In *prokaryotic organisms* where no nuclear membrane exists it is possible for mRNA to associate with ribosomes before the mRNA chain has been completed so that *polypeptide* synthesis can commence. This *coupled transcription* and translation are shown in Figure. In eukaryotes, owing to the cellular compartmentation, the mRNAs must migrate out of the nucleus to the cytoplasm where the ribosomes active in protein synthesis are found. The other main differences between *prokaryotic* and eukaryotic mRNAs are

1. Many prokaryotic mRNAs are transcripts of several adjacent genes (these mRNAs are called *polygenic* or *polycistronic mRNAs),* whereas all eukaryotic mRNAs are transcripts of single genes (these mRNAs are called *monogenic* or *monocistronic mRNAs*).
2. Prokaryotic mRNAs are not processed and have short lifetimes, whereas eukaryotic mRNAs are processed and exhibit a range of lifetimes.
3. Most of the prokaryotic genome codes for mRNA, whereas only a small fraction of the *eukaryotic genome* may code for mRNA. At any one time, less than 10-20% of the total *cellular RNA* in either prokaryotic or eukaryotic cells is present as mRNA.

When the mRNA population of a prokaryotic or eukaryotic-cell is examined, it is found to be heterogeneous in size, with a fairly wide range of S values. This results from the fact that the number of nucleotide pairs in a gene is related to the size of the protein for which they code. However, since a mRNA molecule contains *variable-length* sequences other than those that specify the amino acid sequence of a protein, this size relationship is not absolutely direct.

The general structure of a mature mRNA molecule is shown in Figure.

All mRNAs are transcribed from *structural genes*, that is genes that code for proteins. The three main parts of am mRNA molecule are

1. *Leader sequence* at the 5' end. The length of the leader sequence varies from mRNA to mRNA, and within this sequence is the

information that the ribosome uses to initiate protein synthesis at the correct position. The leader sequence is not translated.

2. *A coding sequence* following the leader sequence. The coding sequence is translated into the amino acid sequence of a polypeptide.

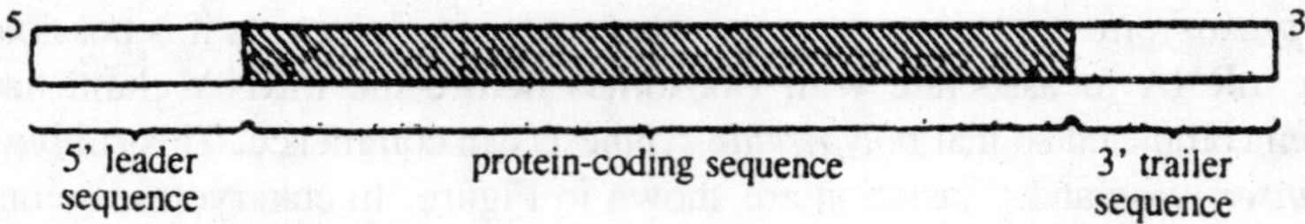

Figure 7.7: General structure of mature prokaryotic and eukaryotic mRNA molecules.

3. *A trailer sequence* at the 3' end. The trailer sequence is not translated and varies in length from mRNA to mRNA. Note that the polycistronic mRNAs in prokaryotes have leader and trailer sequences and spacer sequences between the coding sequences.

Prokaryotic mRNA

In prokaryotes there is generally *colinearity* between a protein-coding gene and the mRNA for which it codes. That is, there is a contiguous sequence of *nucleotide* pairs in a gene and this sequence is transcribed into a contiguous sequence of RNA nucleotides, which are then translated to give the corresponding *contiguous sequence* of amino acids in the protein.

We will see later that this is often not the case in eukaryotic mRNAs. Prokaryotic mRNAs have relatively short lifetimes ; they are quite labile. By this we mean that a newly synthesised mRNA molecule may only function for two or so minutes in organisms such as *E. coli*. After this time, which is approximately 10% of a cell cycle, the RNA will be destroyed.

Thus to continue to synthesize a particular protein, the gene for that protein must be transcribed continuously. This affords the cell a very rapid means of conserving resources because, if a protein is no longer needed for a cell to function, the gene can be blocked from being transcribed and any extent mRNA will soon be destroyed.

This allows the organism to respond rapidly to changing needs when its external environment changes by turning off the transcription of some genes and turning on the transcription of a different set of genes. While most prokaryotic mRNAs are short-lived, some mRNA species in prokaryotes have relatively long lives.

These appear to be resistant to the nucleases that are responsible for the rapid destruction of the short-lived mRNAs. In *E. coil* for example, at least five *ribonuclease* (RNase) activities have been demonstrated, and some of these, if not all, may be involved in mRNA breakdown.

Eukaryotic mRNA

Longevity of mRNAs

Eukaryotic mRNAs are generally long-lived. Compared with prokaryotic mRNAs, many mRNAs in eukaryotic organisms are functional for only an hour or so, which is on the order of 5—10% of the cell cycle in actively dividing cells. These mRNAs could be considered short-lived in these terms, in much the same way as were the short-lived prokaryotic mRNAs. In general, the lifetimes of eukaryotic mRNAs, when expressed as a fraction of the normal cell cycle time, are highly varied.

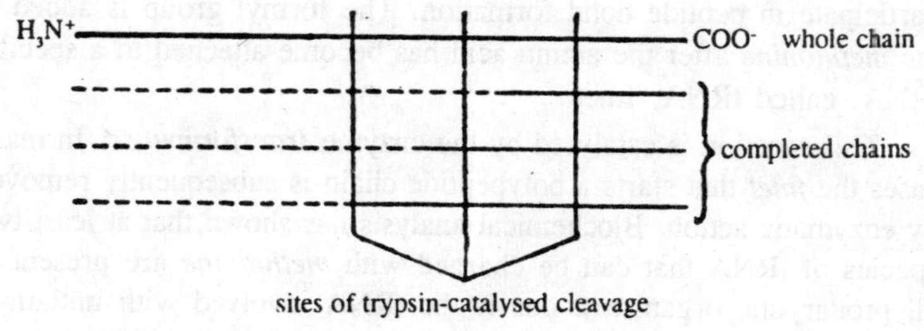

Figure 7.8: Demonstration that polypeptides are made in the N-terminal to C-terminal direction.

This is a practical consequence of the functional characteristics of a eukaryotic cell as compared with a prokaryotic cell. In particular, eukaryotic organisms are differentiated into various cell typeş that have different functions. For example, some cell types are specialised to produce one type of protein almost entirely. Characteristically, the mRNA for that protein will be stable.

Chromatin structure and transcription

As discussed in the DNA of eukaryotic nuclear chromosomes is organised into nucleosome structures. A number of studies have shown that *transcriptionally* active chromatin is more sensitive to *deoxyribonuclease* attack than is nontranscribing chromatin. This result indicates that the transcribed segments of the chromatin must have a different structural conformation than the inactive *chromatin*. However, this is not due to any major alteration in *nucleosome* structure.

5' -and 3' -end modifications

Unlike prokaryotic mRNAs, eukaryotic mRNAs are modified at both the 5' and 3' ends. The modifications occur post-transcription ally and involve the action of specific enzymes.

Protein synthesis in Prokaryotes

There are three major steps in-protein synthesis: *initiation*, *elongation*, and *termination*. These will be discussed in turn.

Initiation

Initiator tRNA

The first amino acid in the synthesis of all bacterial poly-peptides is *N-formyl-methionine* (fmet), which is a modified methionine amino acid in which the α-amino group is "*blocked*" and therefore cannot participate in peptide bond formation. The formyl group is added to the *methionine* after the amino acid has become attached to a specific tRNA, called tRNA. fmet.

This reaction is catalysed by the enzyme *Iransformyiase*. In many cases the *fmet* that starts a polypeptide chain is subsequently removed by enzymatic action. Biochemical analysis has shown that at least two species of tRNA that can be charged with *methionine* are present in all prokaryotic organisms: one is the tRNA involved with initiation, and the other species is responsible for the insertion of methionine elsewhere in the polypeptide chain.

The latter tRNA is designated tRNA fmet. Both of these tRNAs in. bacteria are amino-acylated by the same enzyme, but only tRNA. fmet is a substrate for the transformylase-catalysed: reaction. Both tRNAs read AUG (the only methionine codon),. but in addition tRNA. fmet can recognise GUG and UUG codons. RNA-sequencing studies have shown that both molecules have an anticodon that is complementary to AUG.

The two tRNA molecules do differ in some other properties. For example, the binding of fmet-tRNA. fmet to ribosomes is catalysed by an initiator factor (discussed later), whereas the binding of met-tRNA. met is catalysed by *elongation factors*. The two tRNAs apparently bind to the ribosome at different sites. Thus it is clear that fmet. tRNA. fmet must have a structure that is specific for its role in initiation.

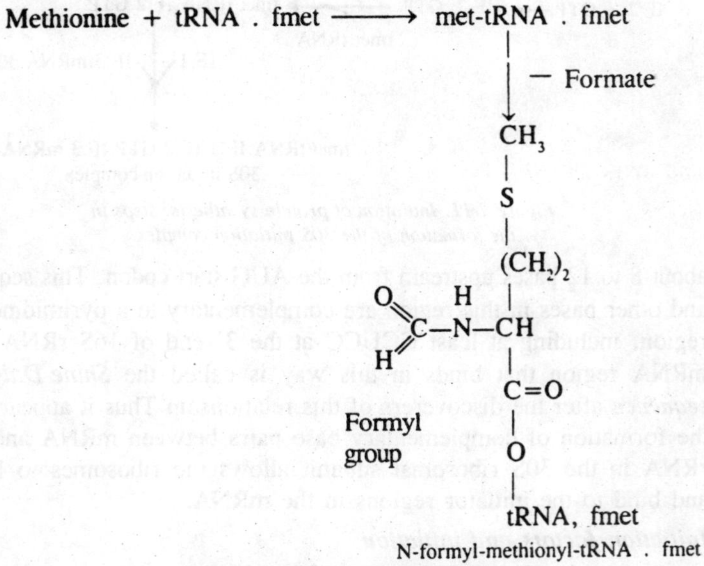

$$\text{Methionine} + \text{tRNA . fmet} \longrightarrow \text{met-tRNA . fmet} \xrightarrow{-\text{ Formate}} \text{N-formyl-methionyl-tRNA . fmet}$$

$$CH_3-S-(CH_2)_2-CH(NH-CHO)-C(=O)-O-\text{tRNA, fmet}$$

Formyl group

N-formyl-methionyl-tRNA . fmet

Figure 7.9: Synthesis of N-formyl-methionyl-tRNA.

Ribosome binding sites

In bacteria, the first step in initiation is the formation of a complex between the 30S ribosomal subunit, fmet-tRNA, and an mRNA molecule. The 50S subunit is added later to form the active 70S ribosome (monosome). The mRNA may contain information for one to several distinct polypeptide chains. For each of the segments coding for a polypeptide, there is a specific nucleotide sequence for orienting the mRNA correctly and in the right reading frame on the ribosome. These sequences are called the *ribosome binding site*.

Message origin	***Ribosome binding site sequence***
E.coli lac Z	UUC ACA C[AG GA]A ACA GCU [AUG] ACC AUG AUU
E.coli trp B	AUA U[UA AGG A]AA GGA ACA [AUG] ACA ACA UUA
E.coli RNA polymerase β	AGC GAG CU[G AGG] AAC CCU [AUG] GUU UAC UCC
Phage λ cro	AUG UAC [UAA GGA GGU] UGU [AUG] GAA CAA CGC

Figure 7.10: Some prokaryotic ribosome binding sites. The initiation codon, AUG, is box 1. The larger boxed regions indicate the regions of contiguous complementarily (including allowable G-U base pairs) to the 3' end of 16S rRNA.

Figure shows the sequence of some prokaryotic ribosome binding sites. Most of the ribosome binding sites have a purine-rich sequence

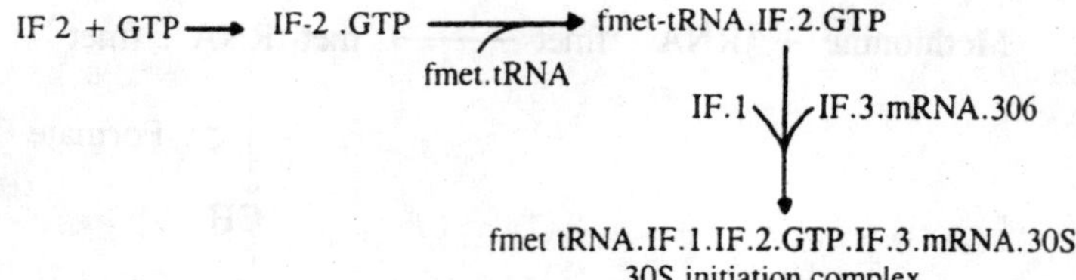

Figure 7.11: Initiation of protein sysnthesis: steps in the formation of the 30S initiation complex.

about 8 to 12 bases upstream from the AUG start codon. This sequence and other bases in this region are complementary to a pyrimidine-rich region, including at least CCUCC at the 3' end of 16S rRNA. The mRNA region that binds in this way is called the *Shine-Dalgarno sequence* after the discoverers of this relationship Thus it appears that the formation of complementary base pairs between mRNA and 16S rRNA in the 30S ribosomal subunit allows the ribosomes to locate and bind to the initiator regions in the mRNA.

Initiation factors and initiation

In addition to mRNA, fmet-tRNA, and ribosomal subunits, ;three protein *initiation factors* (IF-l, IF-2, and IF-3) and GTP are required for the initiation process to occur. First the properties of the initiation factors are discussed and then the scheme proposed for the initiation process in protein synthesis is presented.

1. **IF-3.** The IF-3 factor weighs 23,000 daltons and functions in binding mRNA to the 30S subunit. It also acts as a dissociation factor for separating the 30S and 50S subunits after polypeptide synthesis is complete. Like all the IFs, IF-3 is found bound to free 30S subunits and can be released by washing the subunits in 0.5 M ammonium chloride.

Experiments with radioactive IF-3 have shown that it is capable of binding to both 30S subunits and to mRNA molecules. In an in vitro protein-synthesizing system, IF-3 enhances the binding of fmet-tRNA to mRNA. 30S subunit complexes. It is attractive to suppose that IF-3 recognises mRNAs by the AUG or GUG initiation codons, but there is no solid evidence on this point.

In summary, the initiation reaction in which IF-3 is involved is:

IF-3 + mRNA + 30S subunit → (IF-3. mRNA. 30S) complex

2. IF-2. The 80,000 dalton IF-2 protein is involved with the binding of the initiator tRNA to the IF-3. mRNA. 30S :complex. The high-energy molecule GTP is used in this reaction. In vitro experiments have shown that IF-2 and GTP wilt' bind to form a complex that is stabilised

when it is turn forms a complex with fmet-tRNA. This latter complex then binds with the IF-3. mRNA. 30S complex and the IF-1 protein factor (9000 daltons) to form the 30S initiation complex.

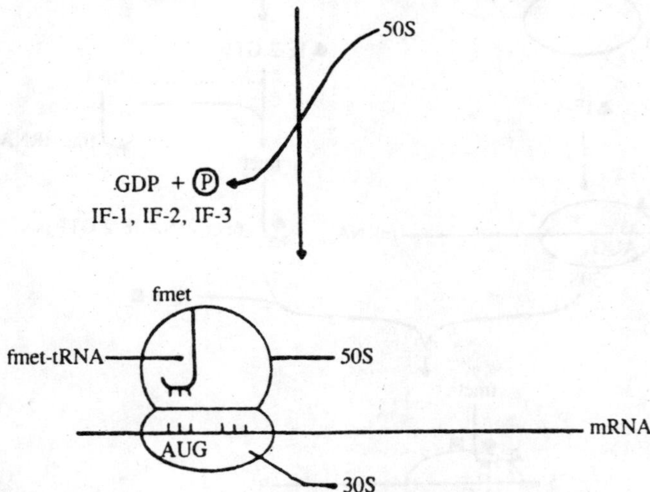

Figure 7.12: Initiation of protein synthesis: addition of 50S ribosomal subunits to 30S initiation complex leads to formation of 70S ribosome in frame on the mRNA.

3. *Dissociation of initiation factors from the initiation complex.* The initiation factors function to bring fmet-tRNA, mRNA and 30S subunits into a stable association. The next step is the addition of a 50S subunit to form a 70S initiation complex. This leads to the hydrolysis of GTP to GDP+P and the release of three initiation factors. The factors can then be used for further initiation reactions on the same or different mRNA.

Elongation

The 70S ribosome has two sites for binding *aminoacyl tRNA*. In protein synthesis, charged tRNA binds first to a site called the A (aminoacyl) site. Then the amino acid it carries becomes joined to the growing polypeptide chain carried by the tRNA at the site called the P (peptidyl) site by the formation of a peptide bond.

It is not known whether the fmet-tRNA enters the A site and then moves to the P site or whether it enters the P site directly. Before further protein synthesis can occur, however, the fmet-tRNA must become located in the P site hydrogen-bonded to the start codon on the mRNA. Once this has occurred a cyclic sequence of events commences in which one amino acid at a time is added to the growing polypeptide

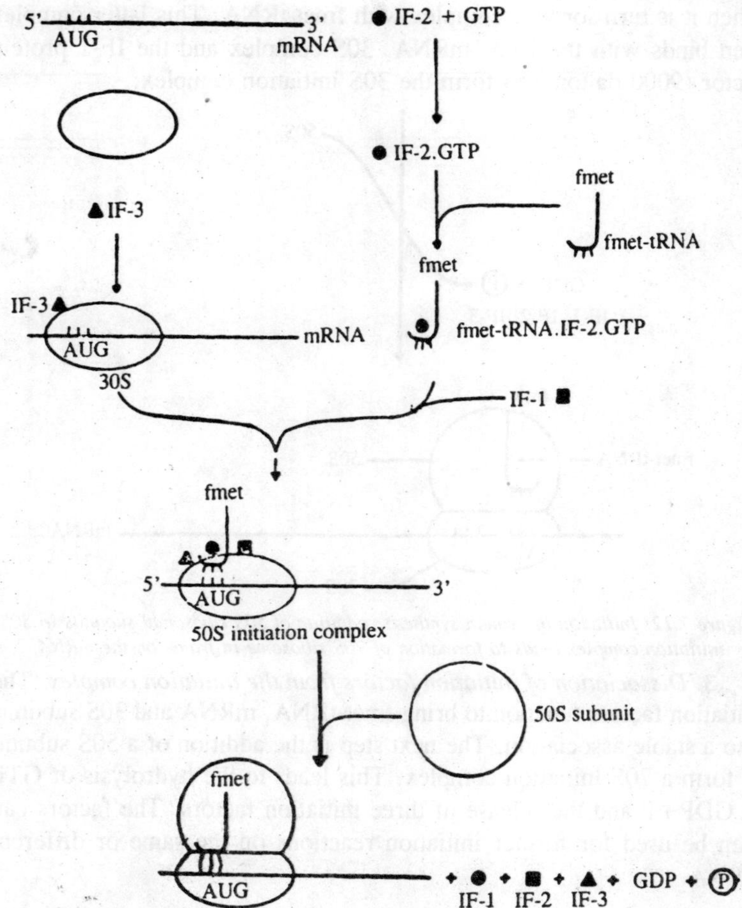

Figure 7.13: Summary of the steps in the initiation of protein sysnthesis in prokaryotes.

chain. This is called *elongation.* These steps will now be discussed in more detail.

Binding of aminoacyl-tRNA

The charged tRNA with the complementary anticodon to the codon in the reading frame of the A site becomes bound to that site of the ribosome in a reaction requiring *elongation factors* T (EF-T) and GTP. This factor can be isolated from the soluble proteins of E. *coli,* and by column *chromatography* it can be separated into two polypeptides: Ts, which is stable and weights about 30,000 daltons, and Tu, which is unstable and weights 42,000 daltons.

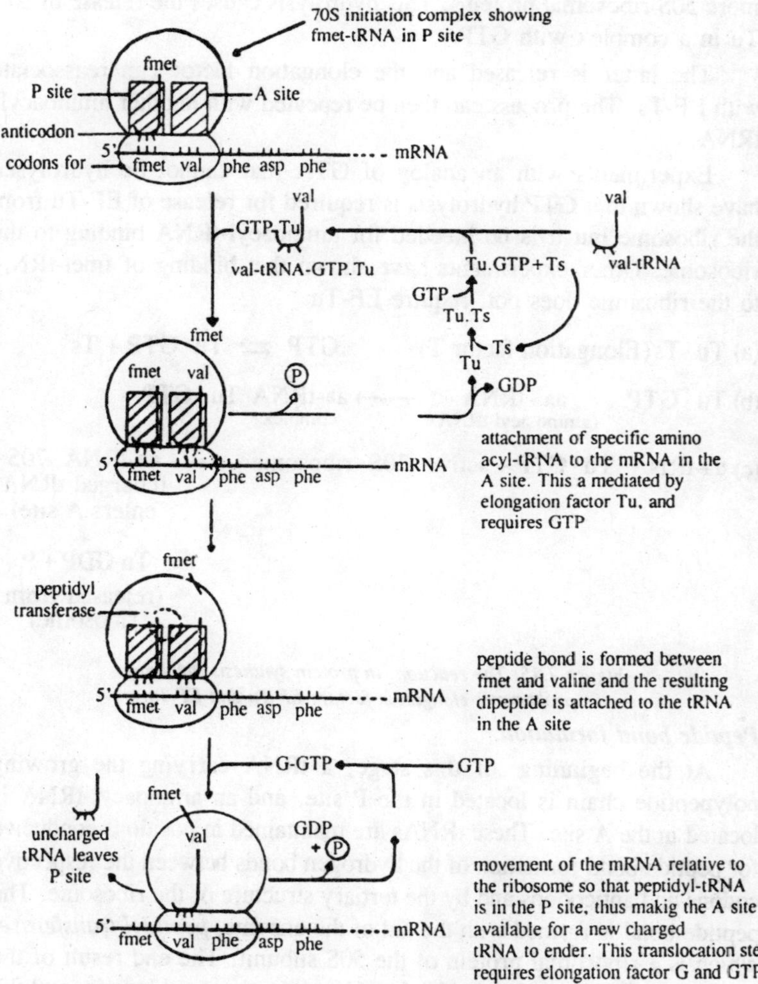

Figure 7.14: Summary of the elongation (peptide bond formation) and translocation steps in protein synthesis.

EF-T has been shown to bind with GTP and this is postulated to bring about the dissociation of the factor into the two polypeptides, resulting in the formation of an EF-Tu. GTP complex and releasing free EF-Ts. The next step in the elongation process is the binding of aminoacyl-tRNA to the complex to produce an aminoacyl-tRNA. Tu. GTP complex.

There is evidence that this complex is an intermediate in aminoacyl-tRNA binding to ribosomes. Once the charged tRNA is bound in the A site, GTP is hydrolysed as a result of the enzymatic action of one or more 50S ribosomal proteins. This hydrolysis causes the release of EF-Tu in a complex with GTP.

The latter is released and the elongation factor can reassociate with EF-Ts. The process can then be repeated with another aminoacyl-tRNA.

Experiments with an analog of GTA that cannot be hydrolysed have shown that GTP hydrolysis is required for release of EF-Tu from the ribosome but it is not needed for aminoacyl-tRNA binding to the ribosome. Other experiments have shown that binding of fmet-tRNA to the ribosome does not. require EF-Tu.

(a) Tu · Ts (Elongation factor T) + GTP ⇄ Tu · GTP + Ts

(b) Tu · GTP + aa · tRNA (amino acyl-tRNA) ⟶ aa-tRNA· Tu · GTP complex

(c) aa-tRNA · Tu · GTP + active 70S ribosome ⟶ aa-tRNA · 70S (charged tRNA enters A site) + Tu·GDP + P_i (released from ribosome)

Figure 7.15: The reactions in protein synthesis involving the prokaryotic elongation factors EF-Tu and EF-Ts.

Peptide bond formation

At the beginning .of this stage, a tRNA carrying the growing polypeptide chain is located in the P site, and an aminoacyl-tRNA is located in the A site. These tRNAs are maintained in positions conducive for peptide bond formation of the hydrogen bonds between the respective codons and anticodons and by the tertiary structure of the ribosome. The peptide bond is formed with the aid of the enzyme, *peptidyl transferase*, which is a ribosomal protein of the 50S subunit. The end result of the reaction is that the polypeptide chain is one amino acid longer, and the growing polypeptide chain has been transferred from the tRNA in the P site to the tRNA in the A site. The tRNA in the P site, which now has no amino acid bound to it, is called an *uncharged* tRNA.

Translocation

Once the peptide bond has been formed and the polypeptide chain is on the tRNA in the A site, the next step is advancement of the ribosome precisely one codon (three nucleotides) clown the mRNA, a process called *translocation*. During this translocation event, the peptidyl-tRNA remains attached to the mRNA by *codon-anticodon-pairing*

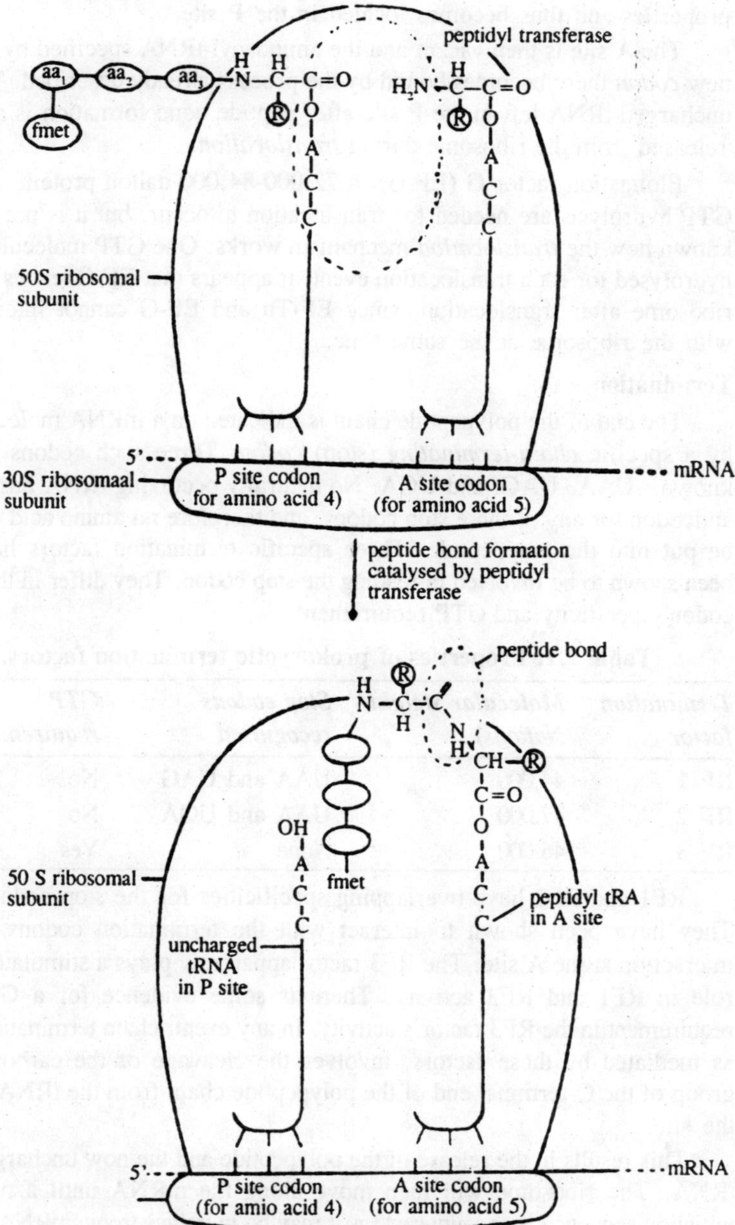

Figure 7.16: Diagrammatic representation of peptide bond formation on ribosomes atalyzed by peptidyl transferase.

properties and thus becomes located in the P site.

The A site is then vacant and the aminoacyl-tRNA specified by the new *codon* there becomes bound by the process already described. The uncharged tRNA left in the P site after peptide bond formation is also released from the ribosome during *translocation*.

Elongation factor G (EF-G), a 72,000-84,000 dalton protein, and GTP hydrolysis are needed for translocation to occur, but it is not yet known how the *translocation* mechanism works. One GTP molecule is hydrolysed for each translocation event. It appears that EF-G leaves the ribosome after translocation, since EF-Tu and EF-G cannot interact with the ribosome at the same time.

Termination

The end of the polypeptide chain is indicated on a mRNA molecule by a specific *chain-terminating* (stop) *codon*. Three such codons are known : UAA, UAG, and UGA. No naturally occurring tRNA has an anticodon for any of these stop codons, and therefore no amino acid can be put into the polypeptide. Three specific termination factors have been shown to be Involved in reading the stop codon. They differ in their codon -specificity and GTP requirement.

Table 7.1: Properties of prokaryotic termination factors.

Termination factor	*Molecular weight Nations)*	*Stop codons recognised*	*GTP requirement*
RF 1	44,000	UAA and UAG	No
RF 2	47.000	UAA and UGA	No
RF 3	46,000	None	Yes

RFl and RF2 have overlapping specificities for the stop codons. They have been shown to interact with the termination codons by interaction at the A site. The RF3 factor apparently plays a stimulatory role in RFl and RF2 activity. There is some evidence for a GTP requirement in the RF3 factor's activity. In any event, chain termination, as mediated by these factors, involves the cleavage of the carboxyl group of the C-terminal end of the polypeptide chain from the tRNA in the site.

This results in the release of the polypeptide and the now uncharged tRNA. The ribosome will then move along the mRNA until a new initiation sequence is encountered (as it may be in polycistronic mRNAs), or it will dissociate from the mRNA. If none is found, when the ribosome is released from the mRNA. IF-3 functions to keep the two subunits apart.

Thus, when a new 70S initiation complex is formed, the two subunits are drawn randomly from the free pools of 30S and 50S sub-units. While the polypeptide chain is being synthesized, the primary sequence of amino acids directs the three-dimensional shape. In other words, the elongating chain begins to assume its final shape as it is being made. Indeed, some enzyme activity can be detected on ribosomes that have not yet completed the synthesis of an enzymatic polypeptide.

Polysomes

Efficient translation of an mRNA molecule cannot be achieved by a single ribosome moving along it. The amount of space a ribosome takes up on a mRNA is relatively small, and, thus several ribosomes can work on the mRNA at once.

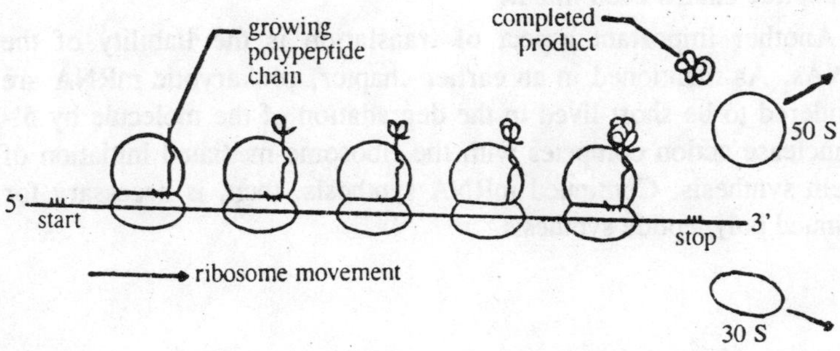

Figure 7.17: Diagrammatic representation of a polysome engaged in protein synthesis.

The association of a number of ribosomes on a single mRNA chain is called a *polyribosome* or *polysome,* and this allows several polypeptide chains to be made from each mRNA. The length of the polypeptide chain on a given ribosome will be directly proportional to how far the ribosome has moved along the mRNA from the 5' end of the molecule. The existence of polysomes explains why a cell needs so little mRNA, while at the same time it contains so much more protein.

Relationship of Transcription and Translation

In bacteria the mRNA typically becomes associated with ribosomes while synthesis of the mRNA molecule is continuing. This is possible owing to the lack of a nuclear membrane so that as the 5' end of the growing mRNA molecule is displaced from the DNA as the double helix reforms, the ribosome binding site becomes available. Ribosomes then load onto the mRNA in rapid sequence, the first being close behind the RNA polymerase.

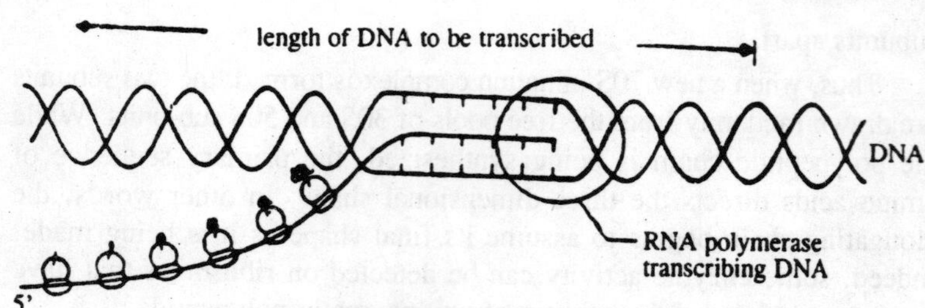

7.18: Schematic of the possible translation of an mRNA while it is still being transcribed in prokaryotes.

To give some idea of the rates of these processes, the mRNA for the tryptophan biosynthetic operon is transcribed at a rate of about 100 nucleotides per minute and the translation process proceeds at about the same rate. Thus approximately 350 amino acids can be polymerised into polypeptide chains each minute.

Another important aspect of translation is the liability of the mRNAs. As mentioned in an earlier chapter, prokaryotic mRNA are considered to be short-lived in the degradation of the molecule by 5'-exonuclease action competes with the ribosome-mediated initiation of protein synthesis. Continued mRNA synthesis, then, is necessary for continued polypeptide synthesis.

INDEX

C

D

E